MODERNITY AND THE MUSLIM WORLD

BY

MOHAMMAD AKRAM GILL

Bloomington, IN   Milton Keynes, UK

authorHOUSE®

AuthorHouse™
1663 Liberty Drive, Suite 200
Bloomington, IN 47403
www.authorhouse.com
Phone: 1-800-839-8640

AuthorHouse™ UK Ltd.
500 Avebury Boulevard
Central Milton Keynes, MK9 2BE
www.authorhouse.co.uk
Phone: 08001974150

First published by AuthorHouse 9/28/2006

ISBN: 1-4259-5672-6 (e)
ISBN: 1-4259-5671-8 (sc)

Library of Congress Control Number: 2006907281

Printed in the United States of America
Bloomington, Indiana

This book is printed on acid-free paper.

Dedication

This book is dedicated to my wife, Fauzia; my sons, Salman, Faheem and his wife Mehvash, and Shuaib; and my daughter, Farheen.

Acknowledgment

A couple of chapters were read by Dr. Khalid Sohail, psychotherapist, Whitby, Ontario, Canada, who provided some general comments and encouragement for publishing the book.

My daughter, Farheen, helped me assemble the manuscript on the floppy disk for submittal to AuthorHouse.

Jeremy Fasbinder of AuthorHouse supported and nudged me on.

I acknowledge their assistance from the bottom of my heart.

Mohammad Gill

PREFACE

The book consists of fourteen essays written between the years 2002 and 2006. Twelve of them were published at chowk.com, one at pakistanlink.com, and one, the first chapter on modernity and the Muslim world, was written specifically for this book. Due to topical similarity between some of the essays, some repetition of the source material and inherent arguments was inevitable, and no effort was made to eliminate it. It is hoped that this repetition is not jarring and may in fact be useful in maintaining the continuity of each essay independently.

The motivation for collecting these essays into a book arose from the realization that books of this kind are rare if not altogether nonexistent. Hopefully, it is useful to draw the attention of the Muslim readers to the causes that are responsible for keeping the Muslim world in a static condition that is becoming increasingly anachronistic and retrogressive. Self-criticism and self-analysis is not the practice of the Muslim world with the result that its doctrines, values, and guiding principles have largely become outdated. It is hoped that this book will instill a desire in the minds of the Muslim readers to change this decadent trend and induce them to march with the changing times to narrow the gap that exists in the scientific and technological spheres between it and the developed nations of the world.

April 2, 2006   Mohammad Gill

CONTENTS

INTRODUCTION

When I was at college, I used to wonder why there was not a single Muslim scientist who made any significant contribution to the science that was taught to us. There was Newton, whose laws of motion are the backbone of classical physics, and many other scientists who applied his laws to solve practical problems. There were others who made original contributions and advanced the frontiers of science. They included Maxwell, Boltzmann, Poincare, Max Planck, Einstein, Niels Bohr, Heisenberg, Schrodinger, Dirac, and many others who made fundamental contributions to the development of modern physics. Darwin heralded a big revolution in biology by propounding his theory of evolution and natural selection. There is not a single Muslim chemist (excepting Ahmed Zewail recently) who made any contribution to modern chemistry although the word "chemistry" is derived from the Arabic al-Kemia.

There was not a single reference to any Muslim scientist in any textbook. No Muslim engineer made any contribution in modern engineering and technology. All the science and engineering textbooks that we used in college were written by foreign authors, most of whom were European and American. These thoughts nagged me all the time. In due time, I tried to understand the reasons for the absence of the science culture in Pakistan and the Muslim world.

I read the history of Arab (Muslim) science. Muslim contributions in science were the hallmark of scientific knowledge up to the time of Copernicus, and there was then darkness afterwards. Muslim contributions in the Middle Ages were conspicuously significant; it may

1

not be completely wrong to say that they were the only people in the world who were doing science. They were the transmitters of science and philosophy to the West. The irony of the whole episode is that I learned all this from books authored mostly by non-Muslim and Western authors. Most of the Muslim scientists are unaware of the contributions that their forefathers made in science. The names of great scientists (e.g., al-Razi, al-Haitham, Ibn-e-Sina, Ibn-al-Nafis, Ibn-e-Rushd, al-Battani, al-Tusi, Ibn-al-Shatir, among numerous others) became familiar to me from non-Muslim sources; their contributions were not included in any of the textbooks at school and college in Pakistan. This knowledge was not part of any school curriculum. We learned science and engineering mostly as part of the foreign knowledge and only as much of it as we could use in practice to earn our livelihood. There was no incentive at any stage for advancing the frontiers of scientific knowledge.

Out of this frustration emanated my desire to write essays related to the history of science to provide some background knowledge to Muslim readers. Many such essays were published at the website, chowk. com. The readers responded in a positive manner with the result that space for a regular column on science (*Science and Sciencibility*) was provided to me by the Chowk administration.

I was in due time introduced to *Sci-Tech* magazine of the Pakistani newspaper, *Dawn*, by a Chowk reader and author, Samina Wahid Perozani, who was also an assistant editor of *Sci-Tech* magazine. I published some essays in that magazine and received a positive response from the Pakistani readership. This inspired me to collect and publish some of these essays in the form of a book to communicate to a worldwide readership. The result of this desire and effort is here in the form of *Modernity and the Muslim World*. If it excites a degree of interest in the minds of the readers—the Muslim readers—and inspires them to contribute to science and inspires others to cultivate the culture of science and modernity in the Muslim world, it will give me a great deal of satisfaction.

Belleville, Michigan
April 2, 2006

CHAPTER 1

MODERNITY AND THE MUSLIM WORLD

> For the Ash'arites (as also for the Sufis), the world is annihilated and recreated at every moment; the cause of all events is the Creator and not a finite, created agent. A stone falls because God makes it fall, not because of the nature of the stone or because it is impelled by an external force. What appears as a "Law of Nature," i.e., the uniformity of sequence of cause and effect, is only a matter of habit, determined by the will of God and given the status of "law" by Him (Nasr).[7]

Yet the trajectory of the fall of the stone can be accurately predicted by scientific formulation without invoking God or any other divine agency. Invocation of God in science is completely redundant; science is not possible with God's intervention.

There are great impediments at the intellectual and conceptual levels for an Islamic society to become modern. This is due to the fact that a predominant majority of Muslim intellectuals are metaphysicians. They recognize the developments that the physical and material sciences have experienced after enlightenment, and almost all of them may be using the products of modern science and technology in their daily lives, but they regard them as subservient and secondary to religious metaphysics. Consequently, whenever a conflict arises between a religious dogma and a scientific fact, it is the dogma that will prevail at all costs. Accordingly, the church, so to say, will always be right and Galileo(s) wrong.

In the practical world, metaphysics doesn't work; it pertains mostly to the next world. It doesn't provide any concrete and practical guidance in this world; whatever it does is motivated by the desire to improve life after death. According to Rudolf Carnap, "This (anti-metaphysical) thesis asserts that metaphysical propositions—like lyrical verses—have only the expressive function, but no representative function. Metaphysical propositions are neither true nor false, because they assert nothing; they contain neither knowledge nor error; they lie completely outside the field of knowledge."[1] Metaphysics is subjective; it varies from one metaphysician to another. Its assertions are vague and can be interpreted in any which way.

For example, consider Nasr's assertion: "We have seen that the sacred art of Islam is an abstract art, combining flexibility of line with emphasis on the archetype, and on the use of regular geometrical figures interlaced with one another. Herein one can already see why mathematics was to make such a strong appeal to the Muslims; its abstract nature furnished the bridge that Muslims were seeking between multiplicity and unity. It provided a fitting texture of symbols for the universe—symbols that were like keys to open the cosmic text."[7] This piece of text contains beautiful words and attractive expressions, but on the whole, it is meaningless. This inanity is characteristic of almost all the metaphysical assertions. And unfortunately, in the Muslim world, the metaphysicians are regarded as "best of the best."

Nasr quotes Omar Khyyam and gives four categories of the "seekers of knowledge," which are as follows:

1. The theologians, who become content with disputation and "satisfying" proofs, and consider this much knowledge of the Creator (excellent is His name) as sufficient.
2. The philosophers and learned men (of Greek inspiration) who use rational arguments and seek to know the laws of logic, and are never content merely with "satisfying" arguments. But they too cannot remain faithful to the conditions of logic and become helpless with it.

3. The Ismailis (a branch of shia Islam) and others who say that the way of knowledge is none other than receiving information from a learned and credible informant; for in reasoning about the knowledge of the Creator, His Essence and Attributes, there is much difficulty; the reasoning power of the opponents and the intelligence (of those who struggle against the final authority of revelation, and of those who fully accept it) is stupefied and helpless before it. Therefore, they say, it is better to seek knowledge from the words of a sincere person.

4. The Sufis who do not seek knowledge by meditation or discursive thinking, but by purgation of their inner being and the purifying of their disposition. They cleanse the rational soul of the impurities of nature and bodily form, until it becomes pure substance. It then comes face to face with the spiritual world, so that the forms of that world become truly reflected in it, without doubt or ambiguities.[7]

This is nothing but metaphysical mumbo jumbo.

Nasr then mentioned, "It is puzzling at first sight to find nowhere in it the mathematicians, of whom Khyyam himself was such an eminent example."[7] The bigger puzzle is, however, that there is no mention of any modern science also (e.g., physics, quantum mechanics, chemistry, biochemistry, biology, astronomy, cosmology, and the whole lot). Nasr, who is an eminent philosopher and Islamic metaphysician and commands great respect both in the Islamic and the Western world, seems to be living in the past when such ideas were formulated. For him, modern science doesn't seem to deserve any specific mention in the categories of knowledge. Science is the dominant part of modern knowledge, and whoever ignores this fact is not being realistic. He believes that the theologians are still the supreme exponents of human knowledge. He is the greatest philosopher of the modern times in the Muslim world after Iqbal, and he didn't mention the modern sciences even incidentally in the categories of human knowledge. How can there be any hope for modernity in Islam?

The peaks of Nasr's and Iqbal's academic attainments were their PhD's in philosophy, which they garnered from USA and Germany respectively. Afterwards, they submerged themselves in the sea of

5

religious metaphysics. Nasr chose a Shiite foundation for his metaphysics (because he is a Shia by birth), and Iqbal built on Sunni Islam. Both of them appear to reject modern science or accord to it only a subordinate position vis-à-vis metaphysics.

Extolling Ghazali's rejection of the Greek philosophy propounded by al-Farabi and Ibn-al-Sina, Nasr asserted, "The famous treatises of al-Ghazzali, in the fifth/eleventh century against the rationalistic philosophers of the time, mark the final triumph of intellection over independent ratiocination, a triumph that did not utterly destroy rationalistic philosophy, but did make it subordinate to gnosis."[7] Unfortunately, that was also the beginning of the end of creative thought in the Muslim world.

Ibn Taiimiyya was a successor and true follower of Ghazali. According to Serajul Haque, "He (Ibn Taimiyya) considers the syllogistic process of thinking artificial and useless. In his opinion, God endowed human beings with necessary knowledge to understand their Creator and His attributes. But men invented, from the very early times, various sciences which the Shariah of Islam does not require for the guidance of mankind…He hates Aristotle and his followers for believing in the eternity of the world (qidam-al-alam), though most of the philosophers were against this view. Ibn Taimiyya rejects this theory (theory of atoms) on the ground that it is an innovation and that early Muslims knew nothing about it. Further, the theologians are not unanimous; some of them totally deny the existence of atoms and the composition of bodies from them…None of the companions of prophet nor their successors nor any one prior to them in natural religion (din-al-fitrah) ever spoke about indivisible atoms."[9] This quotation clearly shows the inadequacy and shallowness of religious belief versus scientific method. A scientific theory is not falsified because the religious belief does not endorse it; only empirical evidence falsifies a scientific theory.

Maryam Jameelah (a convert to Islam from Judaism whose given name was Margaret Marcus), who was a close associate of Maulana Abu-al-Ala Maudoodi, wrote several books on Islam. She seems to write with Maudoodi's authority, and many readers take her quite seriously. She wrote in her *Islam and Modern Man*, "What the Muslim world today needs above all is a modern al-Ghazali and a modern Ibn Taiimiyya. The task of their successors would not be nearly so complicated as it may seem at first because the secular humanism of ancient Greece does not at all essentially differ from contemporary materialist philosophy. The latter is but a further development of the former. One of the most important tasks of our Ibn Taiimiyya is to refute the bogey of progress. Our obsession with 'change' and 'progress' and 'moving with the times to meet the challenge of the age' is nothing but a modern dogma derived from the Darwinian theory of evolution and incorporated into social philosophy as materialist concept of history by Karl Marx. As Muslims, we should be concerned only with submission to the will of God through unquestioning obedience to Quran and Sunnah in its plain, literal meaning." This prescription is nothing if not for the ruin of economic development and economic independence.

The Islamic world is intellectually dominated by the theologians and metaphysicians of the type of Jameelah and Maudoodi; it is their word, which carries the weight and conviction. There is no reputable scientist in the Muslim world that underlines the fundamental importance and relevance of science for the Muslim world. Science, unlike religion, doesn't pretend to present the complete worldview needed by mankind. Science should be integrated with cultural and religious values wherever possible without any coercive enforcement of any particular religion or sectarian or religious belief. Science and religion should cease to remain mutually exclusive and arraigned against each other, which is only possible if religion becomes personal and adjustment between the two is kept at an individual and personal level. Adjustment must not be forced; it should be sought through interpretation. If an individual loses faith and chooses to become an atheist, so be it. It wouldn't be the end of the world.

The Muslim critics are afraid to raise critical issues orally and in print. They are afraid that they might be accused of blasphemy and be treated harshly. Religious criticism is not tolerated in the Muslim world—period. Hiding his identity under the pseudonym of Falsafay Ghaalib,[2] a Muslim writer with a PhD in Economics, wrote, "If according to Muslims, all answers can be found in the Quran, then why did Muslim scholars fail to invent and innovate in modern-day mathematics, physics, chemistry, biology, and interrelated disciplines? Without a single exception, all of these developments have been taking place in the western world, at least in the last two centuries. Second, why did Islam fail to provide rational answers, acceptable to the majority of Muslims to the social and ethical issues resulting from these scientific developments? (I'm thinking of issues like family planning, organ transplant, and cloning and genome research)...The present state of intolerance in the Islamic world was best lamented by a Kuwaiti professor Ahmed-al-Baghdadi. Writing in a local newspaper, he states, 'Muslim(s) claim that their religion is a religion of tolerance, but they show no tolerance for those who oppose their opinions...The Islamic world and the Arab world are the only places in which intellectuals—whose only crime was to write—rot in prison.'...For a healthy society to function, it has to adapt according to changing needs...Social, economic, political and scientific developments are evolving and continuously changing. That is what happened to Muslim societies. Their intellectual curiosity stands frozen in the past millennium."[5]

Having quoted specimens of Nasr's metaphysical writings above, allow me to quote from Iqbal to also show that his metaphysics was of the same genre. Placing "heart" (mind) in a superior position to human reason, Iqbal asserted, "The heart is a kind of inner intuition or insight which in the beautiful words of Rumi, feeds on the rays of the sun and brings us into contact with aspects of Reality other than those open to sense-perception...We must not however regard it (heart) as a mysterious special faculty; it is rather a mode of dealing with Reality in which sensation, in the physiological sense of the word, does not play any part."[3] This seems to be a good piece of poetical prose but what does it actually mean? The body-mind dualism had already been discussed to exhaustion before Iqbal's time and was dying. Perhaps, Iqbal thought he could infuse a new life in it by his poetical inscription underlining the fundamentality of the 'heart.' But for what purpose?

What is Modernity?

Hundreds of books have been written on this subject. To keep our discussion simple and focused, we can start with a historical perspective. History is divided by some writers into three epochs: Antiquity, the Middle Ages, and Modern. According to Wikipedia, "It (Modern) is also applied specifically to the period beginning somewhere between 1870 and 1910 through the present, and even more specifically to the 1910-1960 period."[11] It is conceded that the specification of time periods is largely arbitrary; still modernity can be understood fairly accurately within these specifications. It may also be stated that this specification relates to the developments in the Western world, and this is how modernity is generally perceived, not only in the West, but also almost all over the world.

However, still more appropriate to our discussion here are the "signs of modernity" used by Jene M. Porter. According to Porter, "..other signs of modernity, one could cite the separation of church from state, of power from authority, of individual from society, and of reason from revelation."[8]

The problem that the Muslim world is confronted with regarding modernity is accurately summed up in these signs. All of these four signs are repugnant to the Muslims because of the tight hold that religion has on their minds and because of the pervasiveness of religion in the society. As a rebuttal to the first sign, many will assert with a straight face that the question of separation of church from the state is irrelevant because there is no formal and organized church in Islam. While this is true, this sign needs to be understood as "separation of state from religion" in the context of the Muslim world. The state should be free to choose from the Islamic traditions and practices whatever seems suitable for the constitution of a modern Muslim state without coercion and compelling from any quarters. Religion should be relegated to individual practice. A constitution cannot be crafted for "common good" unless religious compulsion and coerciveness is removed from it.

Another sign from which the Muslim world has not been able to eschew itself is the separation of revelation and reason. The other two signs will be fulfilled easily and naturally from the implementation of these two signs. Science and human knowledge should not be held hostage to revelation. When all the natural phenomena can be, and are, explained in terms of laws of nature, the assumption of a Creator who is the cause of all these phenomena becomes redundant. If one has to invoke a god, he can do so by believing that laws of nature are created by God, but He does not change them, in as much as we know. So, miracles do not happen.

Compared with the metaphysical enunciations, which are difficult to comprehend and are amenable to multiple and diverse interpretations, the rational specifications are easy to understand. Since they can be understood, they can be practiced, also. Practice of the metaphysical mantras uniformly, on the other hand, is difficult since they are vague, subjective, and individualistic, non-uniform in methodology and results, and also quite uncertain. According to Porter, "Reason becomes epitomized by mathematics and geometry. Knowledge is applied to only what can be gained through a realistic method and held with certainty. Here the rational and autonomous individual stands free of authority and tradition."[8] Earth doesn't remain at the center of the universe simply because the Bible said so; it becomes an ordinary planet because the empirical evidence shows it's a planet. Human life didn't begin with Adam and Eve whom God created in His own image some 10,000 years ago because it evolved from a lower form of life over a period of millions of years. The concept of atoms did not need to be discarded because the companions of the holy prophet didn't say that they existed but should have been retained as an unverified concept, which would probably be verified one way or the other in due time. And it has now been verified that they do exist. Unless one is totally oblivious of these realities and chooses to shut his (her) eyes on these and other numerous facts, reason cannot be held subordinate to revelation.

Unfortunately, the efforts are directed in the Muslim world not toward modernizing the society and culture but toward Islamizing science and modernity. Stenberg noted (see ch. 2), "To them the question concerns the Islamization of the modernity."[10] As long as the Muslim world remains intellectually alienated from the West, there is no hope of modernization. The prerequisite for modernizing the society is the acquisition of the knowledge of science from the West where it presently resides. Acquiring knowledge from the West isn't the same thing as blindly following the West in all walks of life; it is quite normal and healthy to practice the Islamic cultural values without their coercive enforcement. The majority of modern Muslims (both practicing and non-practicing) living in the West have a distinctly Islamic culture.

The great difficulty in such an endeavor is that some of the most influential ulema do not want any part of the Western culture. Maryam Jameelah had a long series of correspondence with the renowned Muslim conservative scholar, Maulana Maudoodi. In her seventh letter dated April 12, 1961, she wrote, "I think that the modern-educated ruling elite in Asia and Africa are so obsessed with the mania to develop their respective countries not because they really care about the personal welfare of the poor but rather because they are ashamed of them! They shudder with the most acute inferiority complex every time their countries are labeled 'backward.' I think the mania for industrialization stems not from any real positive benefit the country would gain but because big factories and huge dams and hydroelectric plants would increase their prestige and respect on the part of the 'advanced' countries. Nations are no different from individuals who strain every nerve to accumulate as much wealth as they can just so they can show it off and boast about it. The Holy Quran sums up this attitude beautifully when it says: 'Know that the life of this world is only play and idle amusement, pomp and mutual boasting and multiplying in rivalry among yourselves, riches and children.'"[5]

This outlook is as superficial as the one being criticized. The dams and hydroelectric projects are not built only for national prestige but basically for national well-being. Even if they are built for the wrong reasons, as Jameelah asserted, they are important for economic growth. There is a severe famine in Niger this year (2005), and people are dying of hunger. Who is trying to rescue them from this natural calamity? It is the infidel Western countries that pooled their material resources and provided the much-needed help and relief. How can a poor nation, like any Muslim country without the bounty of oil, help in this kind of situation when it itself is poor? Or should we, the people of conscience, have allowed them to die of hunger because it was the will of God? We should develop our countries in order to at least survive such a calamity, if for no other reasons.

Responding to Jameelah's concern, Maudoodi wrote in the eighth letter on May 19, 1961, "Of course, it is natural that the so called 'underdeveloped' countries should want to put an end to their backwardness as quickly as possible and catch up with the western countries in the race for material progress. But the tragedy is that aid from rich countries is bringing a deluge of western culture in its wake which is a deadly menace to our religion, our morality, our civilization and culture—in short, every thing near and dear to us which makes our lives worth living. Furthermore, the leadership in Muslim countries is in the hands of those persons whose minds are completely vanquished and who venture to re-interpret the laws of the Shariah despite their meager knowledge. Such a situation is doubly dangerous. It not only poses a threat to Islamic patterns of thought and behavior but also there is every possibility of Muslim countries falling into the lap of Communism."[6] In essence, the paranoia of Western culture is holding us back from modernizing and developing our countries.

Jameelah also wrote in her seventh letter, "On Fridays, I go to Columbia University to meet with a group of Muslim students from various countries (including Pakistan) where we gather for Jumaa prayers and then a meal with discussion but as much as my parents', their views clash with mine on almost everything. They firmly believe that Islam must be reconciled with modern western civilization and its ideals and practices modified accordingly. Some even criticize fundamental Islamic doctrines. Many doubt the authenticity of Hadith. Although I make every effort to be polite and tactful, I cannot convince them and they cannot convince me. I always leave with a feeling of frustration."[5]

In his response, Maudoodi wrote in his eighth letter, "The question you asked me at the end of your last letter is an important question indeed! This is indeed precisely the question, which I have been trying to solve for the last thirty-five years. I began my efforts towards understanding Islam and working for its revival when I was a youth of twenty-three and ever since then, I have dedicated my whole life for this task. I never had any faith in mere defensive tactics or a rear-guard action. I have launched a three-pronged offensive. On the one hand, I have ruthlessly attacked the ideological foundations of western culture. On the other hand, I have expounded as fully as I know how, the ideological bases of Islam. I have explained at great length what an Islamic way of life (is) and how in every respect it is superior to western ways (mean). Thirdly, I have offered practical Islamic solutions of important problems, which previously even observant Muslims could see no alternative but to follow the west. As a result of this work, there are millions of Muslims in Pakistan and India from every walk of life who share with me the zeal and yearning for an Islamic order."[6] It seems that Maudoodi's efforts were a reaction to the ever-increasing ubiquity of the western culture; they were not inspired by any fundamental spirit to reform existing Islam, per se. Reactionary work is usually unproductive.

The sad irony in the end was that Maudoodi came to the West for his medical treatment and died in an American hospital. He had condemned the West all his Life, and in the end, he showed that he had greater confidence in the Western medical doctors and its hospitals than in the Muslim doctors and hospitals. Surely many of his followers were doctors and physicians.

Rhetoric is not the same thing as realism. The ulema like Maudoodi have unwittingly misled the Muslim world in as much as modernization is concerned with the best intentions in the world.

References

1. 1. Carnap, R. Philosophy and Logical Syntax—Chapter on "The Rejection of Metaphysics,] " 1935, http://nb.vse.cz/-sloukova/FIL.418/carnap.htm.
2. Ghaalib, Falsafay. *"The Quest for Modernity in Islam,"* http://www.freerepublic.com/focus/news/678771/posts.
3. Iqbal, Allamah Muhammad. *The Reconstruction of Religious Thought in Islam.* Lahore, Pakistan: Sh. Muhammad Ashraf, Publishers, Booksellers, and Exporters, 1999, p.15.
4. Jameelah, Maryam. *"Islam and Modern Man: The Prospects of an Islamic Renaissance,"* http://www.irfi.org/articles/articles_251_islam_and_modern_man.htm.
5. Jameelah, Maryam. "Seventh Letter—April 12, 1961," http://www.cocg.org/booksl/other/maryamj/maryamj_7.htm
6. Maudoodi, Abul-al-Ala. "Eighth Letter—May 19, 1961," http://www.cocg/booksl/other/maryamj/maryamj_8.htm
7. Nasr, Seyyed Hossein. "Science and Civilization in Islam," http://www.fordham.edu/halsall/med/nasr.html.
8. Porter, Jene M. "Modernity: Its Nature and the Causes of its Growth," http://www.artsci.Isu.edu/voeglin/EVS/Jene%20Porter.htm.
9. Haque, Serajul, "Ibn-Taimiyya.in *A History of Muslim Philosophy,"* M.M. Sharif, ed. Delhi: Low Price Publications, 110052, p. 810, 814.
10. *Stenberg, Leif. "The Islamization of Science or the Marginalization of Islam: The Position of Seyyed Hossein Nasr and Zia uddin Sardar,"* http://www.hf.uif/smi/paj/Stenberg.html.
11. Wikipedia, "Modernity," http://en.wikipedia.org/wiki/Modernity.

CHAPTER 2

WHAT IS ISLAMIZATION OF SCIENCE?

Some eight or ten years back (1995-97), I came across some articles on Islamization of science. Some of them were authored by Seyyed Hossein Nasr, an Iranian contemporary philosopher who commands respect all over the world for his work on the philosophy of Islam. I was naturally attracted to his work, and I wanted to comprehend what exactly were the objectives of those who were engaged in Islamizing science. Nasr's writings were general and usually expressed in scholarly phraseology, but they neither explained succinctly as to what Islamization of science was or how it would be prosecuted. His discourse was foggy and philosophically enigmatic. He somehow wanted to bring science into the fold of Islam (which is nearly impossible because science, at least physical science, is totally different from Islam or any other religion) or put a label of Islam on science, as if by affixing 786 on the forehead of science, the objective of Islamizing will be accomplished (n.1). Another fact that intrigued me was that all the papers and articles that came to my attention were written by the philosophers, sociologists, theologians, etc.; none of them was written by a physical scientist (a physicist, chemist, biologist, molecular biologist, etc.). The writers who apparently didn't have firsthand knowledge and experience of the scientific method were suggesting vague methods of Islamizing science.

I continued with my search for more contributions in the hope that I might come across some work of the physical scientists on this topic, but I failed in my search. My search, however, brought me to Leif Stenberg who had written a PhD thesis on this topic and later published a book, *The Islamization of Science: Four Muslim Positions Developing an Islamic modernity.* I could not find this book.

A short article by Stenberg, however, informed me of the four Muslim scholars whose positions he had considered in his thesis. They were Seyyed Hossein Nasr, Ziauddin Sardar, Ismail Raji al-Faruqi, and a French convert, Maurice Bucaille, who had acquired a great deal of prominence in the Muslim world after writing his book *La Bible, le Coran et la Science.* In spite of my concerted efforts, I could not comprehend the goal of the intellectual campaign of Islamizing science or of the methodology to be used for this purpose. True, the authors had written lengthy papers, but they did not define in clear and comprehensible terms what the Islamizing of science was. I wondered if the goal was to Islamize physical sciences together with social sciences, or what? If they aimed at Islamizing physics, for instance, how would they do it? How will they Islamize the theory of relativity and quantum mechanics? Then there are chemistry, biology, micro-biology, and the whole lot. They did not mention any of these disciplines individually but lumped them together in science. Are they going to leave Newton's laws of motion alone, or are they going to stick an Islamic label on them also and baptize them Islamically? I wondered. There were a whole host of questions that mystified me, and I wanted to find some answers. None of the material that I could lay hands on helped me. So I left Islamization alone and got busy with other things. By the way, I would like to mention that at that time, I did not have much knowledge of post-modernism and the Science Wars that were being waged fiercely particularly in the U.S. I learned about Sokal's hoax after a few years of my first encounter with Islamization of science.

A couple of weeks ago (January 2005), I was looking for a book which I knew would be in my daughter's collection. So I went to her room. I didn't find the book I was looking for but came upon *Islamization of Knowledge,*[1] which was published by the International Institute of Islamic Thought (IIIT). Faruqi, one of the four scholars whom Stenberg had considered, helped found IIIT. When I saw the book, my old dormant interest was rekindled, and I thought I might find some new insights and maybe the answer, which had eluded me for so long, was in this book.

I started browsing the book. Although I learned quite a few new facts, my original question (What is Islamization of science?) remained unanswered. The book consisted of 126 pages, yet it did not describe in clear and specific terms how to Islamize knowledge which includes physical sciences. There was no example of any science which had been Islamized. The description is broad, general, unspecific, and theological. It is not a document of scientific precision and accuracy. The book has not mentioned any physical science by naming it. For example, I didn't find any mentioning of the theory of relativity, quantum mechanics, electromagnetic theory, and to top all of them, there was no mention of the unification of weak and electromagnetic interactions for which Abdus Salam got the Nobel Prize. Before this part of quantum mechanical theory could be justifiably claimed as Islamic science, Salam was excommunicated from Islam. The Muslim theologians balk if anybody calls Salam a Muslim. Also, the prize was shared by Sheldon Glashow and Steven Weinberg, both of whom are ethnically Jewish but atheists.

My further search led me to the book *Crisis in the Muslim Mind*, that exists on the Internet under the umbrella of IIIT. In Chapter V, under "Islamization, Science, and Technology," it stated, "It is therefore extremely important for Muslims to realize that not all of Western knowledge and science is objective in nature. It is not difficult to see how the social sciences are clearly subjective, it should not be difficult to see how the hard sciences are really any different in this respect. If there is a difference, it is one of degree only."[2] This is the stance of post-modernism which our scholars have used for the purpose of Islamization. If the hard sciences are subjective, they can be screwed. But did anybody find the need to determine whether the hard sciences were indeed objective? Did they find different "laws of motion" in the Islamic world from those in England, home of Newton who formulated some of them, or in the Western world? The foundation of the physical sciences is their objectivity. Such a statement does not diminish the physical sciences but does show the ignorance of those who are making it.

Then the author goes on to say, "There is no way to speak truthfully about objectivity in science other than from Islamic perspective."[2] Is that the reason that there are no Muslim scientists of note in the world? May Allah protect us from our own ignorance and the sense of self-righteousness.

What Is Islamization of Science?

Before dwelling on this question in any detail, let me mention that the phrase "Islamization of knowledge" was first used and proposed by the Malaysian scholar Syed Muhammad Naquib al-Attas in his book *Islam and Secularism*.[6]

I thought, to find an answer to this mysterious question, a good starting point should be Stenberg's book. Unfortunately, I couldn't find it; it's out of print and none of my area libraries had a copy in stock. Perforce, I had to depend on reviews of Stenberg's book. In this respect, two sources were particularly helpful. One of them is Stenberg's article, "The Islamization of Science or the Marginalization of Islam,"[3] and the other is Muzaffar Iqbal's review of Stenberg's book.[4]

Discussing the neutrality of science, Stenberg wrote, "Both Sardar and Nasr argue that science is not neutral and that it is western in character. Sardar's conclusion is that science therefore is bound to a certain culture. Therefore, it is also possible to create an Islamic science." If science developed by the Muslim scientists is to be called Islamic science, sure, it can be created if the Muslim scientists concentrate and produce some original work in science. But this sort of symbolization is apocryphal. Science is sometimes called Western because mostly the Westerners (including Christians, Jews, atheists, and others) worked to develop it. Science itself is not inherently so conditioned that it can only be developed by the Westerners. Science is neutral in as much as anybody can develop it. In our times, significant contributions have been made by the Chinese, Japanese, Russians, Indians, and a Pakistani scientist who was denigrated in his own country because he did not belong to the mainstream Islam. Science in itself is without religion (it's secular); it has no nationality and has no gender.

Sardar's perception of science is revealing of probably his leanings toward post-modernism. This is evidenced from the above comments of Stenberg and also in the review comments of Muzaffar Iqbal. He wrote, "For Sardar, science is not an objective phenomenon or activity but a cultural activity. Modern science is seen as too deeply rooted in the western civilization which in turn is seen as a threat to the Muslim culture and civilization."[5]

A great deal of confusion was created in the last few decades of the twentieth century in which the post-modernists had claimed that physical science was not objective as believed by the scientists. It was subjective and culture-based like social sciences. They said physical science was socially constructed. Viewed from this perspective, Islamization can have some meaning although it has already been refuted by the scientists. This view was short-lived and is already dead. For more information, see my paper "Is Physical Science Socially Constructed?" chowk.com, February 21, 2004.

Stenberg further tried to elaborate as to what Islamic science is. He wrote, "In their (Nasr, Sardar, and others) approach, one important task is to establish the true interpretation of the word of Allah in order to live the perfect life in accordance with the Islamic tradition. Science must, therefore, be Islamic. In its correct shape it will reveal the true understanding of nature, and increase our comprehension of the creation. Science has a meaning. To be noted here is that science that is in opposition to the Quran will not be accepted. It is not a good science. Science becomes good automatically when it is in accordance with the Quranic text."[3]

More specifically, Nasr asserted, "Islamic science came into being from a wedding between the spirit that issued from the Quranic revelation and the existing sciences of various civilizations which Islam inherited and which it transmuted through its spiritual power into a new substance, at once different from and continuous with what had existed before it."[5] He didn't define what Islamic science is. His statement is absolutely meaningless. If by science, Nasr meant physical sciences, his view is at the worst wrong and at its best vacuous and misleading. Science is fundamentally secular.

He continued confusing the issue further by stating, "As a matter of fact nothing could be further from the truth, for no idea, theory or doctrine entered the citadel of Islamic thought unless it became Muslimized and integrated into the total worldview of Islam."[5]

This view is so archaic that it is amazing to know that people still entertain it. Quran is not a book of science. You cannot reject a theory of science because Quran does not validate it; it has its validation from empirical evidence. No wonder the Muslims are still in doldrums. There is another point to consider. Science in itself is not good or bad. The humans who use it make it good or bad.

The theological scholars like Nasr are the real cause of the backwardness of the Muslim world. How was the Optics, for example, of al-Haitham Muslimized first, before he formulated it? How is science baptized or Islamized? It simply baffles human comprehension.

Stenberg concluded with the remarks, "At stake for the four voices in the discourse is not the modernization of Islam. To them the question concerns the Islamization of the modernity." It is unfortunate because realistically such a task is nearly impossible.

Even the creationists do a better job of defending Christianity against Darwin's theory of evolution than those who are proclaiming to Islamize science. According to the Muslim theological view expressed above, there is no need to defend religion against the theory of evolution; it is waste of time because theory of evolution is simply not Islamic science. Hence, it is not acceptable, period. Some others have gone to extreme and ridiculous lengths to find support for the theory of evolution in the Quran.

This kind of approach for Islamizing science is not scientific; it is theological. It also does not mean that there is any other scientific approach to Islamize science. Such attempts are totally retrogressive. If the Muslims truly want to advance in science, they should do science. Religion is not science. Those who do science may believe in any religion; it is their personal concern. Belief in a religion is not a prerequisite for doing science. As a matter of fact, some notable scientists who were born Jews, Christians, etc., lost their faith after making noteworthy contributions to science. Dirac was an atheist, and so is Weinberg. Their contributions in quantum mechanics are great and beautiful. By separating religion from science, its (religion's) practical usefulness is in no way diminished. The Muslims should recognize this obvious fact and should move on.

Conclusion

The methodology for Islamizing science as postulated by various Muslim scholars is not clear. Their discourse is muddled and inaccurate. These proposals are at best a theological discourse having no relation to the hard sciences. At their worst, they are the result of confused and emotional but ambitious wishes and desires. At best, they fantasize a utopian adventure leading nowhere.

If the Muslim world accepts that science is a worthwhile pursuit and worthy of acquisition and further development for its own sake, there shouldn't be any need for Islamizing it. The fundamental laws of nature that science has discovered are not characteristic of any religion or culture. The ethics of hard science is different from any of the other disciplines.

Laplace is reported to have presented a copy of his book *Mechanique Celeste* to emperor Napoleon Bonaparte. Napoleon had been informed that Laplace's book contained no mention of God. Napoleon asked Laplace, "They tell me you have written this large book on the system of the universe and have never even mentioned its creator."

"I had no need of that hypothesis," said Laplace. And that is the honest truth.

The fact is that science works fine without the hypothesis of divine intervention. Those who are spending days and nights in doing science know this fact. To bring God into science is a step in the wrong direction.

Science does not have any religion, but the scientists do. Many of them are atheists, also. When Professor Salam learned that he had won the Nobel, he prostrated in his namaz (Islamic ritual prayer) and thanked Allah for honoring him with the award. His science did not have Allah in it.

While the objective of Islamizing science may appear laudable and emotionally uplifting, it is unachievable.

Notes

n.1 The sum of the numerical values of the letters comprising "Bism'Allah Alrahman Alrahim" (In the name of Allah, Most Gracious, Most Merciful) is equal to 786.

References

1. "Islamization of Knowledge: General Principles and Work Plan." *International Institute of Islamic Thought.* Herndon, Virginia, USA, 1989.

2. "Crisis in the Muslim Mind." Yusuf Talal DeLorenzo, trans. The International Institute of Islamic Thought, 1993, http://www.usc.edu/dept/MSA/humanrelations/cris_s_in_the_muslim_mind/ch5.html.

3. Stenberg, Leif. "The Islamization of Science or the Marginalization of Islam: The Position of Seyyed Hossein Nasr and Ziauddin Sardar," http://www.hf.uib/smi/paj/stenberg.html.

4. Iqbal, Muzaffar. Review of *The Islamization of Science: Four Muslim Positions Developing an Islamic Modernity,* http://www.cis-ca.org/reviews/4-pos.html.

5. Nasr, Seyyed Hossein. *Islamic Science: An Illustrated Study.* World of Islam Festival Publishing Company, Ltd., 1976, p.9.

6. Attas, Syed Muhammad Naquib al-, "Islam and Secularism," quoted in Wikipedia, http://en.wikipedia.org/wiki/Islamization_of_knowledge.

(First published at chowk.com on January 19, 2005)

CHAPTER 3

DEATH OF RATIONALISM IN THE MUSLIM WORLD

The vendetta between human intellect and divine revelation (religion, Iqbal's intuition or *ishq*) is quite old; in fact, it dates back to the time when man first started to argue about physical objects and natural phenomena using his reason. In the beginning, man offered simplistic explanations for natural phenomena. For instance, it rains because God wills it. And God has created the sun to provide heat and daylight and the moon for moonlight during the night. Again, that God punishes the evil and the wicked and rewards the good people.

Pointing toward this primitive state of human thought, Popper, described in the context of Greek mythology, "To put it crudely, the pre-scientific myth-makers said, when they saw a thunderstorm approaching: 'oh yes, Zeus is angry.' And when they saw that the sea was rough, they said: 'Poseidon (the Greek god of the sea) is angry.' That was the type of explanation which was found satisfactory before the rationalist tradition introduced new standards of explanation."[5]

With the passage of time, man started to notice certain regularities, order, and periodicities in some natural phenomena (changing of the weather, for instance), and using such observations, he was able to make some simple predictions. After gaining confidence, man continued attempting to explain the happenings around him using his observations and reason. Gradually, he started explaining some natural phenomena

25

which in the past he had ascribed to God. The gradual development of his intellectual faculty gave birth to the intellectual process which we call rationalism.

It is now well understood that the natural phenomena in the material world can be explained rationally without invoking divine facilitation; such of them as have defied explanation so far can possibly be explained in the future when man's knowledge will have appropriately increased for this purpose.

What about the spiritual events and the supernatural entities? Those are beyond the realm of reason because they cannot be sense-perceived. Rational explanations are crucially dependent on the sense-perception data; hence the realm of rationalism is confined to the material world only.

Now the question arises: In what way does the conflict of human intellect (reason) and divine revelation arise? I have mentioned above that in the primitive stage of his development, man could only explain the natural events in terms of some god. When eventually he became capable of providing rational explanations to many natural phenomena and systems and synthesizing the observed facts into hypotheses and theories, some conflicts came to light between the old "divine" and the new rational explanations. The believers of divine explanations wouldn't give up their original simplistic explanations because that seemed to detract some power from their perceived God, who was in most cases believed to be almighty and immutable. That is how the Galileo and Copernicus affairs cropped up.

Although the accounts of the fiercest battles between the Christian revelation and human intellect are now buried in the European history of the seventeenth to nineteenth centuries, and a symbiotic equilibrium now exists between them in the West, the issue is still alive in the Muslim world. After the downfall of Baghdad in 1258 CE and the contemporaneous eclipse of the Abbasid Empire, the orthodox ulema had gained ascendancy in the Muslim world and banished, so to say, the rational tradition, which the great Muslim scientists and philosophers had established. Their battle cry was to return to the roots of Islam and do away with the innovations. The seeds of such orthodox thinking already existed even before the fall of Baghdad.

Ibn Taimiyyah who was born soon after the fall of Baghdad started preaching in his adult life to the Muslims to return to the original roots of Islam in which philosophy and the rational sciences did not play any role. Despite the apparent anachronism, this appealed to the defeatist Muslims who wanted to recapture the old glories. Ibn Taimiyyah despised rationalism and the philosophers and chose to adhere to the literal meanings of the holy scriptures.

Serajul Haque has commented on Ibn Taimiyyah's attitude toward philosophers and sciences in these words: "He (Ibn Taimiyyah) considers the syllogistic process of thinking artificial and useless. In his opinion, God endowed human beings with 'necessary knowledge' to understand their Creator and His attributes. But men invented, from the very early times, various sciences which the Shariah of Islam does not require for the guidance of mankind...He hates Aristotle and his followers for believing in the eternity of the world (*qidam-al-alam*), though most of the philosophers were against this view."[8] He poured out his disdain further as described by Serajul Haque, "Ibn Taimiyyah rejects this theory (theory of atoms) on the ground that it is an innovation and that early Muslims knew nothing about it. Further, the theologians are not unanimous; some of them totally deny the existence of atoms and the composition of bodies from them....None of the companions of Prophet nor their successors nor anyone prior to them in natural religion (*din-al-fitrah*) ever spoke about indivisible atoms. Naturally, therefore, it cannot be suggested that those people ever had in mind the term 'body' and its being an assembly of atoms. No Arab could conceive of the sun, the moon, the sky, the hills, the air, the animals, and the vegetables being combinations of atoms. Was it not impossible for them to conceive of an atom without any dimension? The traditionists, the mystics, and the jurists never thought of such doctrines."[9] According to Serajul Haque, Ibn Taimiyyah also believed, "..Allah has given to the Muslims more knowledge and perspicuity of expression combined with good action and faith than all classes of people."[9] Ibn Taimiyyah seemed to have his facts wrong. According to him, our predecessors and forefathers had greater knowledge than us, and the knowledge that they had was unchanging and permanent The fact is that knowledge is growing with the passage of time.

Unfortunately, he could not realize that human knowledge did not begin with the Arabs, and the ancient Greek philosophers should not be despised simply because they were non-Muslims (they couldn't be Muslims since they lived much earlier than the advent of Islam) or non-Arabs. Human knowledge is built brick by brick. Whosoever ignores the contributions of the ancient Greek philosophers to human knowledge is culpable of committing a grave mistake. The Greek knowledge and their rationalist tradition are the valuable heritage of humankind.

For example, Anaxagoras who was born around 500 BCE (more than one thousand years before the advent of Islam) was a great philosopher. According to Bertrand Russell, "It was he who first explained that the moon shines by reflected light…Anaxagoras gave the correct theory of eclipses…The sun and stars, he said, are fiery stones but we do not feel the heat of the stars because they are too distant."[7]

Ibn Taimiyyah had started his onslaught on the philosophers from where al-Ghazali had left it earlier. For instance, al-Ghazali stated in his religious foreword of his *Incoherence of the Philosophers*, "The sources of their (al-Farabi, Ibn Sina, and others) unbelief is their high-sounding names such as Socrates, Hippocrates, Plato, Aristotle, and their likes, and the exaggeration and misguided-ness of groups of their followers…There is no basis of their unbelief other than traditional, conventional imitation, like the imitation of Jews and Christians since their upbringing and that of their offspring has followed a course other than the religion of Islam."[1]

Al-Ghazali opposed rational thought with a dedicated ferocity. Explaining the nature of fire, he said, "Our opponent claims that the agent of burning is the fire exclusively; this is natural, not a voluntary agent, and cannot abstain from what it is in its nature when it is brought into contact with a receptive substratum. This, we deny, saying: The agent of the burning is God, through His creating the black in the cotton and the disconnection of its parts, and it is God who made the cotton burn and made it ashes either through the intermediation of angels or without intermediation…Indeed the philosophers have no other proof than the observation of the occurrence of the burning, when there is contact with fire, but observation proves only simultaneity, not causation, and in reality, there is no other cause but God."[2]

Although it is spooky, Al-Ghazali's claim that his postulate could not be logically refuted is true. Similarly, he could not prove that it was indeed God who burnt the cotton when it was brought in contact with fire. A statement which is irrefutable is not necessarily true. According to Popper, "There have been thinkers who believed that the truth of a theory may be inferred from its irrefutability. Yet this is an obvious mistake, considering that there may be two incompatible theories, which are equally irrefutable—for example, determinism and indeterminism. Now since two incompatible theories cannot be true, we see from the fact that both theories are irrefutable, that irrefutability cannot entail truth."[6]

Regardless of the above philosophical argument, al-Ghazali's thesis is of the same genre as "Zeus is angry" and "Poseidon is angry." While the Muslim world is still mired in condemning rationalism, thanks to al-Ghazali, Ibn Taimiyyah, and their followers, the rationalist West built on its materialistic understanding of the burning action of fire and, in due time, developed elegant theories of heat, thermodynamics, heat engines, and much else using empirical information together with human intellect rather than totally depending on "divine causes" and "divine explanations." Rationalism opened the doors of scientific development in the West and the negation of rationalism shut those very doors in the Muslim world and pushed it farther into mythological phantasm. The metaphysical influence of al-Ghazali is so far-reaching and grasping in the Muslim world that many of its modern thinkers continue indulging uncritically in the metaphysical arguments in upbraiding human reason and elevating revelation (intuition and the like) over it. One of the most influential modern thinkers of this kind was Allamah Muhammad Iqbal.

Iqbal's Perspective and His Dilemma

Although Iqbal greatly appreciated the contributions to physical sciences of the medieval Muslim scholars, he is generally silent about their rational attitude and proclivities. When it came down to discussing divine revelation (inspiration, intuition, *ishq*, heart, etc.) and human reason, Iqbal invariably downgraded reason. Regarding "heart," he said, "The heart is a kind of inner intuition or insight which in the beautiful words of Rumi, feeds on the rays of the sun and brings us into contact

29

with aspects of Reality other than those open to sense-perception…We must not, however, regard it (heart) as a mysterious special faculty; it is rather a mode of dealing with Reality in which sensation, in the physiological sense of the word, does not play any part."[3] This seems to be a good piece of poetical prose, but what does it actually mean? I'm not sure if I understand it.

To avoid confusion and conflict, he could have allocated different universes (spheres of influence) to reason and revelation in which each of them operated (e.g., material world to reason and the spiritual (non-material) world to revelation), but it appears that he didn't do so. It is true that he confined reason to the material world, but his "intuition" is superior to reason and holds sway everywhere. Even in the material world, Iqbal's intuition is superior to reason. Although he wrote extensively on this subject, he failed to clarify the competing issues; he used a muddled religious metaphysical approach in describing them. His description is usually in very broad, general, and non-specific terms. It is difficult to obtain any specific and practically usable information from his metaphysics. But then, religious metaphysics is seldom clear; it is always shrouded in the mists of vagueness and incertitude.

His poetical works are replete with verses extolling intuition and degrading reason. In his *Fikr-e-Iqbal*, Khalifah Abdul Hakim devoted one whole chapter to *"Aql per Iqbal ki Tanqeed* (Iqbal's Critique of Reason)."[4] The author expressed Iqbal's views, saying, "…Uncertainty is the death of both the individual and the community. For this reason, Iqbal instructed the community to stay away from philosophy because apparently, the faith of the common and uncommon people appears to be already weak. If the thinkers indulged in the wonderment which is the beginning and the end of philosophy, they will not be able to adopt any line of action with strong conviction." This is exactly the reverberation of al-Ghazali's and Ibn Taimiyyah's views.

I now give some of Iqbal's verses on this subject in the following to illustrate what he thought of reason in comparison to his *ishq*, intuition, and heart.

> *Khrid waaqif naheen haiy naik-o-budd sey*
> *Badhi jaati haiy zalim apni hadd sey*
> *Khuda jaanay mujhe kiya ho gaya haiy*
> *Khrid bezaar dil sey, dil khrid sey*

[Reason is not conversant of good and bad
The wretch is exceeding its bounds
God knows what has happened to me
Reason is sick of heart, heart of reason]

Alaaj aatish-e-Rumi kay soz mein haiy tera
Teri khrid peh ghalib haiy farngiyun ka fasoon

[Your cure lies in the heat of Rumi's fire
Your reason is subdued by the magic of the West]

Kheerah nah kar saka mujhe jalwa-e-daanish-e-farang
Surmah haiy meri aankh ka, khak-e-Medina-o-Najaf

[The flash of the West's intellect failed to dazzle me
Because my eyes are lined with the dust of Medina and
Najaf]

Bura nah maan zara azmaa kay dekh issay
Farang dil ki kharabi, khrid ki maamoori

[Don't take ill but test it
West is the sickness of heart and fullness of intellect]

Ilm mein bhi saroor haiy lekan
Yeh woh jannat haiy jiss mein hoor nahin

[Knowledge (symbol of reason) has its own pleasure, however
This is a paradise in which there is no houri]

Bey khatr kood pada aatish-e-Nimrod mein ishq
Aql thi mahu-e-tamashaiy lab-e-baam abhi

[Ishq jumped fearlessly into Nimrod's fire
While reason was still watching from above]

There are hundreds of other similar verses in his poetry.

Iqbal is envious of the material advancement of the West. He wished the Muslim world was not so backward, yet he would not support rationalism which is indeed the key to material development. He is afraid that atheism would accompany the advent of rationalism in the Muslim world.

In one of his verses, he says:
Hum tau samjhay thay keh laigi faraghat taaleem
Kiya khabr thi keh chala aaega alhaad bhi saath

[We believed that education would bring prosperity
We didn't know that atheism would also come along with it]

If one wants to acquire rational knowledge, one knows how to go about it. If a person wants to become a civil engineer, for example, (s)he will go to an engineering college to acquire the required knowledge. Afterwards, (s)he will be able to design bridges and after acquiring sufficient training and practical experience, design dams, airports, etc. If one wants to become a physicist, philosopher, chemist, etc., (s)he knows how to achieve her/his objective. But if one wants to acquire the so-called superior knowledge, say knowledge of intuition, where should (s)he go? And how would (s)he use such knowledge in practical life?

There is a great deal of confusion and conflict in Iqbal's thought. Iqbal and his intuitionists existed in this world, but they aspired, most of the time, to working ceaselessly to attain a distinguished station in the next world. They, so to say, lived in this world for the next world. They had almost turned their backs on the reality of this world and devoted their lives to formulate apocryphal religious metaphysics, which would guide them to secure a select place in the next world. They, like al-Ghazali, devoted more energy to the eschatological premises than to the real problems of the world in which they physically lived. For example, al-Ghazali declared the philosophers (al-Farabi, Ibn Sina, and others) kafir on three questions, one of which related to the resurrection of the dead bodies on doomsday. Al-Ghazali ruled that the dead bodies would be resurrected in the exact forms in which they had died. The philosophers thought it was not possible and that the scriptural description of resurrection was merely a metaphorical discourse. Such philosophers who refuted physical resurrection were to be punished by death, ruled al-Ghazali. This issue is of no importance in the real world that we inhabit. Maintaining good health should be more important in this world than the conjectures as to what would happen to the dead bodies in the next.

They (the intuitionists) condemned rational knowledge and sought esoteric knowledge. There is nothing wrong in this quest if this is really what they wanted. But the trouble is that they wanted much more. They wanted to excel in the pursuit of material knowledge also without attaching much value and credence to the sense-perception information. In short, they wanted to achieve it by impossible means. Their esoteric knowledge is worthless in the natural world. This is the conundrum which they refused to comprehend and confront.

The scientists who engage in research do get intuitive flashes once in a while. Such intuitive directions may not always lead to the correct solution, but the scientists do test them whenever they get such flashes. However, research is mostly hard work. It is sweat and grind. It took three hundred years for the mathematicians, for example, to find a proof of Fermat's Last Theorem.

Can Science Survive?

A skeptical and rational attitude is fundamentally important for doing research. It is very important to ask reasonable questions. No research can be accomplished if there is no problem to work upon. Divine explanation kills the spirit of enquiry and is thus mortally dangerous to scientific research. If all the questions can be answered with the invocation of God, where is the need for doing any research?

So, can science survive in the Muslim milieu? Although the title of this chapter is quite pessimistic and depressing, there is a ray of hope that science would survive and even a scientific tradition might take root in the Muslim world. Rationalism is not completely dead; it is in suspended animation.

However, the reasons for germination of a scientific tradition in the Muslim world are mostly external so far. For example, the incessant enmity between Pakistan and India drove Pakistan to develop its own nuclear infrastructure to deter India which already had developed nuclear devices. A crash program was undertaken in Pakistan for the development of nuclear technology. Technology is not fundamental science, but it derives its lifeblood from science. Some spin-off research facilities may have come into existence, which might facilitate further research in other related areas.

No matter what ulema may rule, for or against scientific research, it will have to continue if Pakistan is to survive as an independent sovereign state. The same kind of dynamic forces are working in the rest of the Muslim world. A theocratic government in Iran is struggling very hard to build its own nuclear infrastructure. Even Saudi Arabia, the conservative of the conservatives, is secretly planning for the acquisition of the nuclear know-how. If the technological infrastructure is established in the Muslim world, scientific research and development together with rationalism cannot be excommunicated. Although the way these developments are taking place at present is not the natural way in which a scientific tradition should be created, it is nonetheless opening up vistas for the growth of science and technology.

References

1. Al-Ghazali. *The Incoherence of the Philosophers*. Michael Marmura, trans. Utah: Brigham Young University, 1997, p.2.
2. Averroe (Ibn Rushd). *The Incoherence of the Incoherence of Philosophers*. Simon van Den Bergh, trans. The Trustees of the E.J.W. Gibb Memorial, 1987, pp. 316-317.
3. Iqbal, Allamah Muhammad. *The Reconstruction of Religious Thought in Islam*. Lahore, Pakistan: Sh. Muhammad Ashraf, Publishers, Booksellers, and Exporters, 1999, p. 15.
4. Hakim, Khalifah Abdul. *Fikr-e-Iqbal*. 2 Nursinghdas, Garden Club Road, Lahore, 1968, pp. 317-318.
5. Popper, K. *Conjectures and Refutations*. London and New York: Routledge Classics, 1963, p. 169.
6. Ibid. p. 264.
7. Russell, B. *A History of Western Philosophy*. New York: Simon and Schuster, 1972, p. 64.
8. Haque, Serajul. "Ibn Taimiyyah.in " *A History of Muslim Philosophy*," M.M. Sharif, ed. Delhi: Low Price Publications—110052, p. 810, 814.
9. Ibid. p. 809.

(First published at chowk.com on December 2, 2004)

CHAPTER 4

METAPHYSICAL OBFUSCATION

> Khwajah Naqshband says that all that is heard or seen
> or known is a veil. It must be negated with the word "none"
> (la).[1]

Dualism of mind and matter (the mind-body problem) goes back to the ancient Greek philosophers. Matter, of which our universe is constituted, was regarded as inferior to mind which was considered a divine quality. A close study of ancient Greek philosophy reveals that many of the world religions are greatly influenced by Greek thought. At the same time, it seems plausible that Greek metaphysics was influenced by the older civilizations such as the Babylonian, Egyptian, and the Hindu civilizations, etc.

According to Russell, "Pythagoras (569-475 BC) taught that...the soul is an immortal thing, and that it is transformed into other kinds of living things; further that whatever comes into existence is born again in that all things that are born with life in them ought to be treated as kindred."[4] He believed in transmigration of the soul (Hindu cycle of rebirth and reincarnation). The premise that the material world is only an illusion (maya) goes back to Pythagoras. Quoting Burnet, Russell described Pythagoras, "We are strangers in this world, and the body is the tomb of the soul, and yet we must not seek to escape by self-murder; for we are the chattels of God who is our herdsman, and without his command we have no right to make our escape."[4] Quoting Cornford,

37

Russell stated of Pythagoras, "All the systems that he (Pythagoras) inspired tend to be otherworldly, putting all value in the unseen unity of God, and condemning the visible world as false and illusive, a turbid medium in which the rays of heavenly light are broken and obscured in mist and darkness."

The Greeks called mind "nous" which was also identified with soul. Discussing Anaxagors (500-428 BC) Russell said, "He (Anaxagoras) differed from his predecessors in regarding mind (nous) as a substance which enters into the composition of living things, and distinguishes them from dead matter...Mind has power over all things that have life; it is infinite and self-ruled, and is mixed with nothing."[5]

Likewise, Plato (427-347 BC) and Aristotle (384-322 BC) believed in the supremacy of mind over matter. A culmination of these metaphysical ideas occurred in the philosophy of Plotinus (204-270 CE) according to whom there is a "...Holy Trinity: The One, Spirit, and Soul...The One is somewhat shadowy. It is sometimes called God, sometimes the Good; it transcends Being, which is the first sequent upon the One...The One can be present without any coming: 'while it is nowhere, nowhere is it not'...The One is indefinable, and in regard to it there is truth in silence than in any words whatever...We now come to the Second Person, whom Plotinus calls 'nous.'"[6] Nous is difficult to translate into English. According to Russell, "Nous, we are told, is the image of the One; it is engendered because the One, in its self-quest, has vision; this seeing is nous...Thus when we are 'divinely possessed and inspired' we see not only nous, but also the one....Soul is the third and the lowest member of the Trinity. Soul, though inferior to nous, is the author of all living things; it made the sun and the moon and stars and the whole visible world."[6]

I have produced this rather extensive excerpt in order to emphasize that Plotinus's ideas are the bedrock as well as the fountain spring of the subsequent metaphysics of most of the Sufis and mystics. Many subsequent metaphysical developments were inspired directly and indirectly by Plotinus's metaphysics. So much so, the state of *wajd* (ecstasy, trance) into which many qawwali listeners are transported, goes back to Greek times. According to Russell, "The experience of 'ecstasy' (standing outside one's own body) happened frequently to Plotinus."[7]

In due time, many of Plotinus's ideas were Islamized so seamlessly that many of the later Muslim theologians and theosophists believed them to arise from Quran. The monotheist concept of Tauheed (Unity of God) is similar to the concept of Plotinus's One. The Islamic Sufistic concept of soul and spirituality has its roots in Plotinus's metaphysics and is equally muddled cognitively.

In spite of its excessive emphasis on the otherworldliness, metaphysics of religion is a branch of epistemology and should be considered as such. It is not the termination of all knowledge; it should not exclude the pursuit of material knowledge as has happened selectively in the Muslim world.

The tragedy in the Muslim world that set it apart from Europe and other growing civilizations was excessive emphasis on metaphysics of religion compared with other branches of knowledge which were practically excluded from its culture. Metaphysics was the only branch of philosophy that was accepted in the Muslim world; other branches, including logic and physical sciences, were denigrated if not totally prohibited. Metaphysics became the epitome of philosophy.

Metaphysics, as we have seen, deals with the knowledge of God, the "Absolute Reality," and the otherworld. It holds knowledge of the material world in low esteem. The knowledge based on and derived from sense-perception is unreliable according to the metaphysics of religion. The ontological knowledge of the Absolute Reality is identified with the nous (Arabic "*nafs*") or the mind, and the sense-perception knowledge is identified with the material knowledge of the world gained through human intellect. The human intellect by the usual definition is lowlier than the mind. Since according to this line of thought, mind is superior to reason, the only worthwhile knowledge is the knowledge of the Absolute Reality.

In this way, the majority of the Islamic thinkers and philosophers got trapped in the otherworldliness. Al-Ghazali condemned the philosophers and found salvation in Sufism. He found his enlightenment only after abandoning philosophy and the pursuit of material knowledge. I have emphasized al-Ghazali's role in directing the Muslim world away from rationalism in other chapters also because he is held in such an inimitable esteem in the Muslim world. He is recognized as the Mujaddid (Revivalist) of the first millennium. He denigrated philosophy and physical sciences but embraced metaphysics passionately. He thus demarcated a definite line of thought for the subsequent theologians and theosophists to follow. Every other form of knowledge was a kind of innovation.

Wahdat-al-Wujud (Existential Unity)

> If man is actually part of God, the evil in man is also in God.
>
> Bertrand Russell[8]

Metaphysics is a boundless ocean of otherworldly concepts. It is difficult to describe it, even in summary form, within the scope and framework of this chapter. Usually, the words used to describe metaphysical concepts are the familiar ones, but when they are strung into sentences and assembled together to describe metaphysical concepts, the whole usually becomes a meaningless jumble which can be interpreted in numerous and sometimes contradictory ways. They are a maze of blind alleys which do not lead anywhere. One such concept is *wahdat-al-wujud*, the existential unity, propounded by ibn-Arabi. Even this concept seems to have its roots in Greek metaphysics. This has been selected herein only for the purpose of illustration. Other concepts of Sufism and mysticism are equally vague and shrouded in the mist of incertitude.

There is a Persian idiom which summarizes the essence of *wahdat-al-wujud*. This is "*Hama O'ast*" (Everything is from Him or everything belongs to Him). According to this thinking, all the living things have unity of being. (Of course, this is not the same thing as Darwin's theory of biological evolution). The culmination of human life is to seek oneness with God or to get merged in Him (*Fana-fi-Allah*). A good example of this kind of thinking was Hallaj who had pronounced "*Ana-al-Haq*" (I am God, or I am one with God) in a "state of deep absorption." He was declared an apostate and hanged.

I am digressing a little bit here to illustrate that metaphysics, pantheism, and mysticism were not the exclusive preserves of Muslim thought only, the Hindus and the Western world also freely indulged in them. However, it is the Muslim world which got inextricably trapped in it. According to Angelus Silesius, a German mystic of the seventeenth century, "God is the fire in me, and I am the light in Him; do we not intimately belong to each other?...I am as rich as God; there is no grain of dust that I (Believe me, O Man) do not have in common with Him...God loves me above Himself; if I love Him above myself I give Him as much as He gives me out of Himself...Without me God cannot make a worm; if I do not preserve it with Him, it must straightaway fall to pieces...I know that without me God cannot live for an instant; if I come to nothing He must needs give up the ghost...,"[10] so on and so forth.

A great follower of ibn-Arabi was the celebrated Maulana Rumi. According to Hitti, "Al-Rumi shares with ibn-Arabi theories of existentialist monism. He identifies himself with nature, following a system of transmigration, and rejoices not in a personal life continuing beyond the grave but in self-integration in the person of Godhead:

I died as a mineral and became a plant,

I died as a plant and rose to animal,

I died as animal and I was man.

Why should I fear? When was I less by dying?

Yet once more I shall die as a man, to soar

With angels blest; but even from angelhood

I must pass on: all except God doth perish.

When I have sacrificed my angel soul,

I shall become what no mind e'er conceived.

Oh, let me not exist! For Non-existence

Proclaims in organ tones, "To Him we shall return."[2]

This is beautiful poetry, and I am very fond of Rumi's poetry. Beyond that, this narration does not mean anything to me. Even if, for the sake of argument, there is a kernel of truth in it, it is unverifiable. I am glad that I did not waste my life in trying ceaselessly to comprehend the incomprehensible (thus going in circles); I was content in contributing, in my own small way, to material knowledge.

In the metaphysical world, quantum mechanics, nuclear physics, Einstein's theory of relativity, etc. are insignificant. These sciences are truly meaningless for those who are bewitched by the beautiful poetry of Rumi and unintelligible webs and woofs of ibn-Arabi and other metaphysicians. The more meaningless a metaphysical description is, the truer it seems. Is there any wonder then that many people, including philosophers, could not truly understand Iqbal's metaphysics but were convinced nonetheless that it was extremely meaningful?

Sheikh Ahmed Sirhandi, Mujaddid Alif Thani (Revivalist of the second millennium) was initially enamored of *wahdat-al-wujud* but later on discerned it was against the teachings of Quran. According to Farman, "I had accepted pantheism, says the Mujaddid, as it was revealed to me and not because I was directed to it by some one else. Now I denounce it because of the right revelation of my own which cannot be denied although it is not compulsory for others to follow..."[1] The Muslims believe that Prophet Muhammad (pbuh) was the last prophet, yet the chain of divine revelations seems to remain unbroken. Mujaddid Alif Thani alluded to his own revelation in the above quotation, and many other Sufis also claim to experience revelations.

Although Mujaddid Alif Thani rejected ibn-Arabi's *wahdat-al-wujud*, later on, Shah Wali Allah claimed to have reconciled Mujaddaid's basis of rejection with the original postulation of ibn-Arabi.[9] Metaphysics is wholly a subjective exercise and "play on words." One can invariably get what one wants out of the metaphysical hat.

Based on *wahdat-al-wujud*, many Sufis deduced that all beings had the same origination, i.e., God, and thus they could possibly merge unto Him. According to the Hindu and Greek concept of transmigration of the soul, the "beings" continue returning to life after death in various forms until they find *mukti* (nirvana, salvation) and merge with God. Since man can and will ultimately merge with God, according to their beliefs, many wujudis claimed that the difference between man and God is only an illusion; in reality, they are the same. Parvez has quoted verses of many Sufi poets, e.g., Bullhe Shah, Khwaja Ghulam Farid, et al., which describe this unity. Many Sufi poets including Iqbal have unified Ahmad (name of the prophet) and Ahad (One—God's attribute) by removing 'm' (Urdu 'meem') from Ahmad. Parvez quoted Iqbal's following verse to illustrate his point:

"Nigah aashiq ki dekh latee haiy purdah-e-meem ko utha kar

Woh bazm-e-Yathrib mein aa kay baithain hazar mun'h ko chhupa chhupa kar"[3]

("The lover's eye removes the veil of 'meem' and gets the vision
Even though He hides his face sitting in Yathrib")
Again, Parvez quoted Bullhe Shah's following verse to make the same point:

Ahad, Ahmad wich farq nah Bullahya

Ikk ratti bhar marodee da

(O Bullhe, there is no real difference between Ahad and Ahmad but that of a small rounded point [i.e. of m or "meem"])

Such conjectural exercises and denigration of the material world and its knowledge through physical sciences ushered in a dark age in the Islamic world in which it is still trapped. Those who want to indulge in metaphysics and mysticism may do so; knowledge of physical sciences should, however, be not degraded, and scientific traditions should be allowed to take roots and germinate in Muslim society.

Finally, I want to quote Parvez, a modernist Islamic scholar and reformer, who has severely criticized tasawwuf. According to him, the qualities and essentials of Sufism are as follows:

- There is a direct relation and communication between man and God. It is called gnosis or "esoteric knowledge" for which no rhyme and reason is needed nor is there any need of proof and evidence.
- Esoteric knowledge is "sure knowledge." Compared to it, the sensible knowledge (knowledge through sense-perception) or intellectual knowledge is of little importance. Sensible knowledge is based on incertitude and analogy. So much so, even the revealed knowledge (wahi) is also not real; it is true only in as much as is certified by the esoteric knowledge.
- The Universe of Senses (phenomenal universe) is not real; it is merely an illusion, superstition, and imagined. The real

existence is only that of God and the phenomenal universe is His exponent.

- The end and objective of human life is for the human mind to merge with the "absolute reality." For this reason, Sufism is the name of totally individual and subjective experiences.
- The more a human draws himself away from the mundane attractions (which are effectively material dirt), the more his spirituality progresses. It evidences itself in the form of predictions and miraculous acts.[3]

I want to conclude this essay with the following substantive excerpt from the essay on mysticism by Bertrand Russell. He is commenting on time which many mystics believe is unreal:

First of all, what can be meant by saying that time is unreal? If we really meant what we say, we must mean that such statements as "this is before that" are mere empty noise, like "twas brillig." If we suppose anything less than this—as for example, that there is a relation between events which puts them in the same order as the relation of earlier and later, but that it is a different relation—we shall not have made any assertion that makes any real change in our outlook. It will be merely like supposing that the *Iliad* was not written by Homer, but by another man of the same name. We have to suppose that there are no "events" at all; there must be only the one vast whole of the universe, embracing whatever is real in the misleading appearance of a temporal procession. There must be nothing in reality corresponding to the apparent distinction between earlier and later events. To say that we are born, and then grow, and then die must be just as false as to say that we die, then grow small, and finally are born. The truth of what seems an individual life is merely the illusory isolation of one element in the timeless and indivisible being of the universe. There is no distinction between improvement and deterioration, no difference between sorrows that end in happiness and happiness that ends in sorrow. If you find a corpse with a dagger in it, it makes no difference whether the man died of the wound or the dagger was plunged in after death. Such a view, if true, puts an end, not only to science, but to prudence, hope, and effort; it is incompatible with worldly wisdom, and—what is more important to religion—with morality.

References

1. Farman, Muhammad. "Shaikh Ahmad Sirhindi." A History of Muslim Philosophy. M.M. Sharif, ed. Delhi: Low Price Publications—110052, 1999, p. 879.

2. Hitti, Philip K. *Islam—A Way of Life*. Chicago: Henry Regnery Company, 1970, pp.62-63.

3. Parvez, Ghulam Ahmad. *Tasawwuf Ki Haqeeqat*. Gulberg, Lahore: Adarah Talu-e-Islam, 1981, p.97.

4. Russell, Bertrand. *A History of Western Philosophy*. Simon and Schuster, 1972, pp. 33, 32.

5. Ibid., pp. 62-63.

6. Ibid., pp. 288-289.

7. bid., p. 290.

8. Russell, Bertrand. "Mysticism," from *Religion and Science*. Oxford University Press, 1961.

9. Siddiqi, Abdul Hamid. *"Renaissance in Indo-Pakistan: Shah Wali Allah Dihlawi."*A History of Muslim Philosophy. M.M. Sharif, ed. Delhi: Low Price Publications—110052, 1999, p. 1570.

10. Steiner, Rudolf. (Archive) *Mysticism at the Dawn of the Modern Age—Giordano Bruno and Angelus Silesius*, http://elib.com/Steiner/Books/GA007/English/GA007_Giordano.htm.

(First published at chowk.com on March 16, 2004)

CHAPTER 5

DECLINE OF SCIENCE IN THE MUSLIM WORLD

> If in the long run scientific thought and intellectual creativity in general are to keep themselves alive and advance into new domains of conquest and creativity, multiple spheres of freedom—what we may call neutral zones—must exist within which large groups of people can pursue their genius free from the censure of political and religious authorities. Insofar as science is concerned, individuals must be conceived to be endowed with reason, the world must be thought to be a rational and consistent whole, and various levels of universal representation, participation, and discourse must be available. It is precisely here that one finds the great weakness of Arabic-Islamic civilization as an incubator of modern science. (Toby E. Huff)

By virtue of patronage at the highest level during the Abbasid caliphate in Baghdad, the foundations of the structures of rational and natural sciences were laid and the House of Wisdom (Bait-el-Hikmah) was formed in which the scholars of all creeds, religions, and beliefs were invited to work. Works of science, philosophy, metaphysics, mathematics, etc. were acquired from the Greek, Indian, and other sources; they were translated into Arabic, the lingua franca of that time, and the scholars were encouraged to advance the frontiers of the existing knowledge further. The result of such consistent efforts emerged in an excellence in science that was unique and unequalled in the previous history of science. According to Huff, "...Up until the thirteenth and fourteenth centuries Arabic science was so developed and promising as to be called the most advanced in the world. In the case of astronomy, it is obvious that this supremacy existed until the mid-sixteenth century, when the astronomical models of Ibn-al-Shatir and the Maragha School were superseded by the new astronomical system of Copernicus."[2] Their excellence in physics, medicine, and mathematics was also noteworthy. The names of al-Haitham, al-Razi, Ibn-al-Sina, Ibn-al-Rushd, and numerous others were latinized to make them familiar (almost household names) in the Western world.

Copernicus seems to be the watershed in the history of science. After his time, the Muslim world was completely eclipsed in science, and the Western world was at the threshold of what is popularly called "the modern science." Never again until recently did any notable scientist, philosopher, mathematician, or anybody else in creative thought and intellectual arts emerge from the Muslim world on the world scene. For example, there is none in quantum mechanics, with the sole exception of Abdus Salam, whose work could be quoted authoritatively. More or less, the same is true of the theory of relativity. Muslim contributors in modern medical science are few and far between (if any, as a matter of fact).

The Muslims all over the world became defensive in as much as the creative and natural sciences were concerned, and at times, they became irrational in minimizing the importance and relevance of these sciences to humanity. They regressed more and more into the religious shell and ascribed the cause of their decline in world power (hence in science and the creative arts also) as due to deviation of the ummah from the straight religious path that Islam had prescribed. This kind of rationalization is irrational, but they tried (and continue to do so) to find reasons and rationale for everything in religious terms. This mode of thinking is part of the collective psyche of the orthodox Muslim world and has existed from the very beginning of Islam.

Causes of Decline

Unfortunately, rational and natural sciences never became an indissoluble part of the Islamic culture even when it was at the apex of such pursuits and engagements. The development of science never became a communal endeavor and responsibility. Compared with the religious law (*fiqh*) and actual practice of the religion, the pursuit of science occupied only a very subservient position. According to Huff, "If we were to construct a status hierarchy, it would start with the legists, the fuqaha, at the top, below them the mutakallimun (religious dialecticians), and below them, the faylasufs, the philosophers cum natural scientists."[2] The natural and rational sciences were considered secular (in the sense of irreligious) and foreign to Islam and were imported from other societies, particularly Greek. Many mutakallimun considered them (the sciences) as innovations in Islam and therefore reprehensible. They were not allowed to develop in their own right for fear that the practitioners will be misled. In the Muslim world, a secular society is unimaginable even now; religion is not personal but societal. Every person is concerned with the beliefs and religious practice of every other person. Secularism is equated to unbelief and atheism, which is intolerable. The Muslim students may be receiving instruction at schools in the theory of biological evolution, but it cannot be discussed in public with any degree of conviction. For any meaningful research in science, the Muslim students still have to go to the Western universities.

Even in the days of glory of Muslim science, it was not formally taught in madrasas. Even the instruction of medical science, which was regarded as the most prestigious science in the medieval Muslim world, was relegated to the hospitals and individual teachers who taught it at their homes. Such teachers would issue *ijazas* (a sort of certificate) at the successful completion of a course, which authorized a recipient to undertake teaching in the respective field. No organized system similar to the modern university system could evolve and persist, and the instruction of sciences remained individualistic. No formal group research was undertaken although such activities could have taken place at the observatories (some of which were excellent and unequalled) and the hospitals. Some of the hospitals were the best in the world. Most of the scientific works (if not all) were written by individual authors. Since the formal teaching of science did not enter the madrasas, which were better organized mainly for legal and theological instruction, science did not become part of the Islamic culture. This alienation between religion (culture) and science still exists. Science had entered the Muslim world from the outside (it did not evolve from within), and it remained as an external entity; it still is. Once it started to decline, it couldn't bounce back and rejuvenate because it didn't have deep roots in the society.

Many analysts describe political and economical factors also that contributed to the decay of science in the Muslim world, but had it taken root (or been allowed to take root) in the Islamic culture, science would surely have overcome the adversity caused by these other factors.

Conflict of Reason and Revelation in the Muslim World

> A leap of faith is not very certain. Even though reason is not perfect but what good is there in replacing reason by religion whose mainstay is blind faith? What guarantee is there that the blind faith is true and better than a rational truth? A society, which banishes reason and rational sciences from its culture becomes a barren society and is doomed to degrade and decay in spite of the supremely divine and unblemished religion it may claim to possess. (Mohammad Gill)[1]

The orthodox Muslims believed that Judaism and Christianity, the other two Abrahamic religions, had degenerated, lost the original divine spirit, and had become obsolete. Their divine books were interpolated and were unable to provide divine guidance. They believed Islam had transplanted them and the Bible and Torah were replaced by the Quran, which is the final word of God. Since Islam was the true religion and its book the true word of God, Prophet Muhammad was the last prophet because there was no need for any other prophet after him. The Quran, the original and unaltered book of God, will continue to provide guidance eternally to humanity. In the face of such pristine and holistic belief, Quran became the ultimate source of human knowledge. Many started believing that whatever does not exist in the Quran is not worth knowing. The modern world is so very different materially and in every other respect from the one that existed fifteen hundred years ago when the Quran was revealed, yet Quran remains the source of all knowledge to the orthodox believers. To many such believers, even the dramatically new concepts like those of quantum mechanics, exist in the book (Quran). The theory of biological evolution is also there in the book. Many normally denigrate the theory of evolution and consider it blasphemous yet when they are in a defensive mood and want to affirm the universality of Quran, they find allegorical language in Quran testifying that the theory is there in the book.

The situation is indeed weird. Those who developed the theories of quantum mechanics, relativity, and biological evolution, never read the Quran (with any seriousness, if any of them read it at all), and the Muslims whose book it is and who recite it day in and day out, do not even understand these theories. Yet, they believe that these theories exist in Quran. The fact that they were discovered by the unbelievers without any help from the holy book is completely lost on them. If these theories existed in the Quran, why did the Muslims fail to discover them and develop them? The discovery is always after the fact.

A clash was bound to happen between the Muslim scientists and philosophers on one side, and the orthodox believers on the other. The orthodox theologians held a trump card to defeat the rationalists; they could declare a certain theory (theory of evolution seems a good candidate), branch of knowledge (philosophy, for example), and a certain scholar (Ibn-al-Sina and al-Farabi, to wit) as blasphemous and invoke the harshest punishment. The scientists and rationalists had no such weapon to counteract the charge of blasphemy. Strangely, when religion is losing its tight hold in many diverse societies, the Muslim world is still governed by punitive doctrines of blasphemy and apostasy.

The religionists normally proclaim that Islam does not prevent development of the rational sciences; in fact, it promotes them. They will quote verse and chapter from the Quran to support their claim. This is nothing but razzmatazz. At an individual level, one can pursue his research career at home, if he can, or, more frequently, abroad where all the facilities are available for his research pursuit. On the communal and national levels, development of science runs into practical problems. With the religious mindset, one wouldn't know what problems can (or should) be undertaken for research because for many problems, the religion would provide answers (mostly wrong) beforehand so that any research venture would become redundant and useless. The other problems which do not seem to have ready answers from religion are labeled useless and wasteful projects.

Unless the rational view is deeply inculcated emphasizing that the natural phenomena occur according to some natural laws (and not by the will of God) and can be understood and explained by discovering these laws, religion will remain a dominant stumbling block. And in the Muslim world, it still is. There is no compelling reason to invoke God in science.

The religionists believe in divine miracles, and those who believe in miracles are unfit to do science. In science, miracles do not happen. Religion and science can coexist peacefully and flourish if science is released from the clutches (fear of blasphemy and apostasy, for example) of religion. The people should be able to discuss, advance by research, and publish works on the theory of biological evolution, for example, in the local research journals and newspapers free from the fear of blasphemy. Darwin's name should not remain despicable and tarred with heresy and not be used to block publication and dissemination of research results. His work should be adjudged on scientific grounds and not on religious grounds. Even if it is a wrong theory, it should be allowed a life of its own.

References

1. Gill, Mohammad Akram. "Skeptical and Counter-Skeptical Trends in Medieval Islam," (chapter 8), chowk.com, November 28, 2002.
2. Huff, Toby, E. *The Rise of Early Modern Science: Islam, China and the West.* Cambridge University Press, p. 72, 2003.

(First Published at chowk.com on September 1, 2005.)

CHAPTER 6

STAGNATION OF CREATIVE THOUGHT IN THE MUSLIM WORLD

The Islamic world is in a vortex of intellectual confusion and chaos. It is difficult to dissipate the forces of this vortex and find a suitable direction towards progress and modernity. The ubiquitous crisis in the Islamic world is regarding the form of government that a Muslim country should have, term of its tenure, and how the succeeding government should come into force after the expiry of the term of the preceding government. All the other problems emanate from this. According to Roedad Khan, "One of the principal causes of the instability of Muslim rule, past and present all over the Islamic world, is the absence of a law of succession, which inevitably led to uncertainty, civil wars, wars of succession through out Islamic history."[3]

Historically, the form of government in all the Muslim countries was autocratic. It varied from benevolent autocracy to tyrannical, depending upon the personal character and disposition of the ruler. Designating an autocrat a "khalifah" (caliph) did not change the fundamental reality. All the rulers were invariably autocrats whether or not they were truly the "vicegerent of God," which the title of "khalifat-Allah" literally means. It is not very clear how the concept of khilafat in this particular sense gained currency in the Islamic Weltanschauung. According to Hamza Alavi, "...The first khalifah, Abu Bakr Siddiq, was called Khalifat-ar-Rasool, i.e., the successor of the prophet; the other three khalifahs after him were called Ameer-ul-Momineen. In the Qur'an, man, representing the whole human race, was called khalifah, i.e., vicegerent on earth. For example, according to thirtieth verse of Surah Baqar, 'Behold, thy Lord said, I will create a vicegerent on earth. They said, Wilt Thou place therein one who will make mischief and shed blood?'"2

And more often than not, they (the autocrats) came into power by violent means, e.g., by killing the previous ruler, by putting him into prison, or by sending him into exile. Peaceful transfer of power was only occasional. There was no formally written constitution ever which could guide the rulers in the art of governance or advise the citizenry of its civil rights, if it ever had any. There were thus no laws and rules or limits which a ruler was to abide by. The plea for not having a written constitution was readily provided by the ulema, whenever they were in good relation with the rulers, asserting that the Qur'an was the ultimate constitution. There was no need for any manmade constitution when the "word of God" was there for guidance. The trouble with this provision was and still remains that there was more than one interpretation of almost every Qur'anic injunction on any given issue. There were sometimes conflicting injunctions in different parts of the book. Also, Qur'an covers extensive ground in an order which is not easy to decipher, discern, and understand, even by well-learned scholars. There is ample room for debate and dispute. On top of that, there is no universally accepted interpretation of the Qur'an in existence. It can provide a broad and general guidance but where more definitive direction is needed, there is no exact guidance. A constitution, on the other hand, is a concise and clearly written document specifically developed for the purpose of just and fair governance. This document is not ultimate and universal in any sense; it can be and should be modified by constitutional means, according to the changing circumstances.

There are four authoritative schools of fiqh, which are outdated because they originated more than a thousand years ago, but still remain effective because nobody has the will or the recognized authority to amend them according to the needs of the time. Commenting on this malaise that afflicted the Islamic world, and still does, C.A. Qadir mentioned, "Blind imitation of the past became the hallmark of the Muslims. The verdicts of Imams and the jurists were accepted more in letter than in spirit. While the jurists and other religious thinkers never claimed infallibility or finality of their legal and theological decisions, the Muslims thought that the last word had been said on the subject and that amendment or departure amounted to sacrilege."[4]

Another difficult aspect about the four schools of fiqh is that they are deeply embedded in religion, which in the meantime has extensively diversified and lacks unity of thought and purpose. The fact is that the Islam of today is not the same as the Islam of say the seventh or eighth century when the four schools of fiqh came into existence. There are many more sects of Islam now than there were ever before. And the followers of these sects have diverse beliefs and views on almost everything, differing substantially from one another.

For example, there are some seventeen Ahle Hadith groups only in Pakistan, which are presently active in politics and jihadi movements.[1] And these are only of the Ahle Hadith persuasion. What of Shiahs, Sunnis, and others? Each one of these groups is clamoring for establishing an Islamic state (theocracy) in Pakistan according to its own perception. None of these groups has any written manifesto outlining the concept of the Islamic state (with the exception of Jamat-i-Islami, perhaps). How an Islamic state, if it ever came into existence, will interact with the rest of the world (both Islamic and non-Islamic) is not clear. I am fairly certain that it is not clear to the proponents of the Islamic state either. One thing that can be concluded from the utterances of the various Islamic groups is that they will continue waging jihad against the rest of the world until they conquer it or perish in the struggle. This kind of anarchic struggle, even if you call it jihad, doesn't make much sense. This is exactly how the Islamic world had always frittered away its energies and resources.

I remember that during the 1960s, in an election campaign in Pakistan, a television interviewer asked a leader of Jamat-i-Islami about his party's political manifesto. The leader had replied that his party would present its manifesto after winning the election. The fact was that they didn't have a manifesto which any significant portion of the public had endorsed or approved. They didn't win the election either.

Since the ulema and their followers are tied to the concept of an Islamic state based on guidance from the Qur'an, their concepts about everything that depends on religion remain static. And unfortunately, everything in the Islamic world depends on religion one way or the other because religion has become so pervasive. That is the reason that the Muslim world is in turmoil, and it has not made any progress in any physical sciences or any other walk of life.

For example, the Muslim contribution to physical sciences, theoretical science, mathematics and mathematical science, medical science, social science, political science, you name it has been absolutely nothing after the twelfth/thirteenth century. Philosophy was banished from the Muslim world by Imam Ghazali (1058-1111). (See, for example, his *The Incoherence of the Philosophers.*) The subsequent philosophers in the Islamic world indulged mainly in mysticism and metaphysics, which emphasized the "otherworldliness" mostly.

The Muslims are stationary in time and space. They do not want to move along with the changing times. They do not want to understand that the world has changed and the time has changed. The tragic thing is that they do not want to break the mental inertia.

The Muslims need to step out of the box and look at everything objectively. They'll see then that the direction they have chosen to march in is really wrong. They are trying to go backwards in time; they are desperately trying to set the time clock back. They have chosen an impossible mission, and in spite of their recurrent failures, they do not give it up. Development of sciences and technology, improving the economic condition of their people and the state by using better economic practices, improving the quality and standard of education programs are not part of any Muslim country's agenda. They emphasize putting the women folk within the household confines and in veil (hudood), abolishing female education, abolishing television, music, and other means of entertainment, making the day-to-day life impossible to live, and all the other negative things which have somehow become an integral part of their vision. The majority of the people will not accept these things willingly; if force is used, it will generate further chaos. And realistically, even the enforcers cannot live up to their own unpractical standards as the Taliban experience showed in Afghanistan. These ultra-orthodox elements need to be enlightened somehow.

Those who have read Western history are aware of how continuous development took place in every sphere of Western life. Take the development of physical science for instance. The West and European scholars took over from where the Arab scientists in the twelfth/

thirteenth century had left off, translated their books from Arabic into the European languages, and continued making progress step by step. In due time, a Copernicus came along who replaced the Ptolemaic concept of the planetary movements with a more convenient and comprehensive system of his own. Galileo followed Copernicus and fought his battle with the Roman Catholic Church, putting the sun at the center of the solar system instead of the earth, which became one of the ordinary solar planets. He lost his battle to the church at that time but won his scientific argument in the long run. He also laid down the foundation of mathematical physics. He was followed by Kepler and later by Newton who introduced a revolution in physics by defining the laws of gravitation. He also gave us calculus. Faraday and Clark Maxwell developed the electromagnetic theory, and Einstein at the beginning of the twentieth century gave us the theory of relativity. A wholly new branch of physics was founded by Niels Bohr, Heisenberg, Erwin Schrodinger, Max Born, Paul Dirac, and others in the early twentieth century, which is called quantum mechanics. Quantum mechanics was made possible by the earlier revolutionary work of Max Planck who formulated that energy was not a continuum as was commonly believed; it was discrete, like matter, in structure. Energy transfer occurred in finite quanta.

What was happening in the Muslim world in the meantime? Absolutely nothing of any consequence in the field of creative thought and rational sciences, excepting wars of succession, quibbling on petty religious issues, and declaring the pursuit of material sciences and philosophy as deviationist activities. Consider the field of political science. While the Muslim world remained entangled and stagnant in the grip of autocratic rulers, the Western world continued making reforms in political thought and system. Some of the early most prominent Western political philosophers were Hobbes, Locke, and Rousseau, among many others. While the Muslim world is still struggling with the concepts of khilafat and Islamic state, the Western world has experimented with several forms of governance and finally developed the liberal form of democratic government in which people are allowed to vote for their candidates. Elections are held at regular intervals of time, and the transfer of government takes place smoothly. There is a constitution to guide the government for its day-to-day operation, and the constitution is easily available to everyone. The constitution is not difficult to read and understand. There are legislative bodies to amend the constitution whenever there is a need for it. There is no room for usurpation of government by the army generals through coup d'etat. Then there are the fine and cultural arts. Islamic civilization did contribute fine specimens of architecture, although the art of architecture was never formalized into a branch of scientific knowledge and inquiry, for study, research, and further development.

In a way, the concept of khilafat existed in the Western countries also under a different label and designation. The English kings believed that they had a divine right to kingship; in other words, the king, like a khalifah, was God's representative on earth, and his subjects owed allegiance to him like an act of faith in God. By virtue of his being the divine ruler, he could and did misuse his powers to his own benefit. According to Bertrand Russell, "...This type (Divine Right of kings) maintained that God had bestowed power on certain persons, and that these persons, or their heirs, constituted the legitimate government, rebellion against which is not only treason, but impiety."[5]

Oliver Cromwell put the divine kingship to an end, and the way was opened for democracy. The Russian tsars likewise also believed in divine kingship. Describing how the Russian tsars ruled their subjects, Margaret Thatcher stated, "...They recognized no property rights except their own, because they treated all their realm as if they owned it...With no private property—above all, no private land—there could be no law, other than tsar's autocratic decrees..."[6] By replacing "tsar" with "khalifah" in the above description, one also gets a glimpse of how the Muslim autocrats ruled their subjects.

When the British colonies of America cast off their colonial yoke, the founding fathers laid down the foundations of a republican form of government, which was defined as the "government of the people, by the people, and for the people." The American form of liberal democracy is probably the best form of governance in the world. Democracy has evolved after experimenting with different forms of government over a period of several centuries. It became possible because there was no mental hindrance in relinquishing one form of government if it was not for the "common good" of the people. Since all of these forms were manmade, it was easy to change them. If the Muslim world accepts mentally and practically the fact that a government needs a written manmade constitution, which may be based on the teachings of the Qur'an but has a flexibility of its own, things could improve. Every Islamic country should have a constitution of its own in keeping with its own individual identity and addressing the needs of its own people, which may have a different mix of diverse religions and beliefs. Creative thought in the Islamic world continues to remain chained; unless religion is separated from statecraft, the current malignance will persist.

The Muslims are in such a condition at present that even to wage jihad, they need Western weapons; without them, they cannot even fight and hope to win. Yet they condemn everything that has originated in the West. People should have no hesitation in espousing good wherever it may come from. Islam is the religion of peace, and that is what the Islamic world should seek.

References

1. Ahmed, Khalid. "The Power of the Ahle Hadith." The *Friday Times*, July 12-19, 2002.

2. Alavi, Hamza. "Ironies of History: Contradictions of the Khilafat Movement," http://ourworld.compusere.com/homepages/sangat/khiltt.html

3. Khan, Roedad. "The View From Pakistan: 'Cry' the beloved Afghanistan." The *Pakistan Nation*, October 29, 2001.

4. Qadir, C.A. "The Dark Age." Chapter LXX, "Decline in the Muslim World." *History of Muslim Philosophy*, Vol.2. M. Sharif, ed. p. 1429, 1999.

5. Russell, Bertrand. *A History of Western Philosophy*. New York: Simon and Schuster, p. 629, 1972.

6. Thatcher, Margaret. *Statecraft*. London: Harper and Collins, p. 72, 2002.

(First published at pakistanlink.com on August 9, 2002)

CHAPTER 7

THE METAPHYSICS OF RELIGION

> This (anti-metaphysical) thesis asserts that metaphysical propositions—like lyrical verses—have only the expressive function, but no representative function. Metaphysical propositions are neither true nor false, because they assert nothing, they contain neither knowledge nor error, they lie completely outside the field of knowledge.[1]

Broadly speaking, sources of human knowledge are rational, revelational, and empirical. These categories are not sharply discriminated from each other; there are transitions between one category and another. For example, the transition between rational and empirical is somewhat quite obvious although the same is not true of revelational and empirical. Metaphysics, although included in rational knowledge, may also be perceived as a transition between rational and revelational. It can also be argued that perhaps there is no transition between revelational and empirical knowledge; they are quite distinct from each other. The mystical mode of knowledge, on the other hand, can either be considered wholly subsumed by metaphysics or totally an independent category.

Rational knowledge is the knowledge according to reason or that which appeals to reason; this is not necessarily verifiable by direct observation or laboratory experimentation, although much of it is. The knowledge obtained from revelation is given by the word of God; it is obtained from the holy scriptures. Almost all of this knowledge is unverifiable. Empirical knowledge is wholly verifiable and testable. This is indeed the most reliable knowledge that man has.

Religion, in the context here, is understood to consist of a system of beliefs having one (or more) supernatural entity or being as the central and fundamental belief. Although, religion in this prescribed sense will be the subject of discussion here, there will, however, be occasions when reference will be made to individual specific religions also, e.g., Christianity, Islam, Hinduism, etc.

Metaphysics is a field of speculation that encompasses almost every branch of human thought. Physical sciences, which are based and reinforced by empirical information, are excluded from the arena of metaphysics because in such sciences, speculation does not play any central role although it does help sometimes in formulating scientific theories. Any theory that is not verified by empirical evidence will not receive acceptance as a theory of physical science. Metaphysics generally deals with notions, concepts, hypotheses, etc., related to the non-material world, i.e., the sphere of mental events, or the astral world. Knowledge of the material world will sooner or later become part of the physical sciences. Looked at another way, physical sciences are constructed on the basis of the information gathered through sense-perception; on the other hand, the point of departure for the metaphysical epistemology is the belief that the sense-perception data are not reliable. The roots of this philosophical dispute extend to the ancient times.

Protagoras (480-411 BC) was probably the first known philosopher who placed nous (mind) above everything else; to him, mind was indeed the Primal Being or God Himself. Plato (427-347 BC) formulated the concept of form, which was equivalent to Protagoras' nous, and the formless matter was subservient to form. Aristotle (384-322 BC) also believed in the supremacy of the form. A whole litany of metaphysical concepts was conceived by the ancients that continued to be worked upon and enhanced by the succeeding philosophers, mystics, and metaphysicians of all kinds. Modern work on the mind-body problem was inspired by Descartes who "...used the soul to explain intelligent brain functions: there is something like an intelligent creature in your brain which sees what needs to be done and then pulls the right strings. Descartes did not concern himself with the question of just how intelligence works in a physical sense, since at that time intelligence was regarded as a purely spiritual property."[2]

Soul

Soul is a central concept in the metaphysics of religion. The ancient Greek philosophers philosophized about soul, and they invariably believed that the soul was non-material although it was incorporated somewhere in the human body. At death, the body dies and decays but the soul is liberated and remains alive; the soul is immortal. The material body was considered a hindrance for the soul to develop its full potential. During his last hours before dying, Socrates (477-399 BC) discussed the nature of the soul with his friends who were there with him. Socrates was not afraid of dying; in a way, he looked forward to dying, because he believed that after death, his soul will be shorn of the material body and will join the company of gods in the next world. He will then be able to continue his intellectual work unhindered without having to satisfy the needs of the material body. He comments, "A soul in this state makes its way to the invisible, which is like itself, the divine and immortal and wise, and arriving there it can be happy, having rid itself of confusion, ignorance, fear, violent desires and the other human ills and as is said of the initiates, truly spend the rest of time with the gods."[5] Later, Plotinus (AD 200-277) expanded on the concept of soul comprehensively, and his metaphysics eventually found its way into Christianity; Islamic thought was also greatly influenced by it. Plotinus believed in a kind of Holy Trinity: The One, Spirit, and Soul. "The One is indefinable, and in regard to it there is more truth in silence than in any words whatever."[6] Russell, describing Plotinus's philosophy, considered Spirit to be equivalent to nous which "...is the image of the One; it is engendered because the One, in its self-quest, has vision; this seeing is nous...Those divinely possessed and inspired have at least the knowledge that they hold some greater thing within them, though they cannot tell what it is; from the movements that stir them and the utterances that come from them they perceive the power, not themselves, that moves them, in the same way it must be, we stand towards the Supreme when we hold nous pure..."[6] According to Russell, Plotinus described, "..Soul, the third and the lowest member of the Trinity, though inferior to nous, is the author of all living things; it made the sun and moon and stars, and the whole visible world. It is the offspring of the Divine Intellect. It is double: there is an inner soul, intent on nous, and another, which faces the external."[6] A materialistic way of thinking cannot make much sense out of the preceding description of soul. According to modern research, the brain performs all the functions that were considered to belong to the mind (soul). According to Furst, "..it was scientific advances of the nineteenth century, most notably the theory of the evolution of species and the principles of conservation of matter and energy, which annihilated Descartes's soul. For if the universe was a closed system, as nineteenth century physics revealed, with the total amount of stuff in it constant, then intelligent action should be explainable on the basis of pieces of matter in motion, without recourse to non-material soul."[2] The trend of modern research in the mind-body problem indicates that it is likely that some time in the future, the duality of the mind and body issue will get resolved by reducing the so-called mental events into the material events. The research in neuroscience and artificial intelligence are converging on this matter. For instance, according to the Type-Identity theory of Smart, Place, and Armstrong, "Just as water is H2O and common salt is NaCl and the temperature of a gas is mean molecular kinetic energy, mental terms like 'believing,' 'desiring,' and 'loving' will be shown to be synonymous with terms that refer to types of neural events, so that some day we shall be able to say 'love' is such and such activity in sector 1704."

Comprehending the Incomprehensible

Almost all the religions have God as a pivotal Being in whom belief is an essential article of faith. In most religions, God remains, largely undefined, over-defined, or sometimes defined so vaguely that such a definition is of little use, with the result that the God of one religion is not the same as of another. Discussing this confusion, Smith remarked, "What, then, is meant by the word 'god'? This is not a simple question. There have been many historical concepts of god, from the anthropomorphic deities of the Greeks to the omnipotent god of Christianity. Some gods are all powerful, all knowing, and all good, while others are not. Some gods are objects of reverence, while others are not. Some gods communicate with man, while others do not. Differences such as these make it impossible to give a detailed description of a god that will encompass every religion—and securing widespread agreement on the meaning of 'god' is a formidable, if not, impossible task."[7] Lowder also dwelled on the same theme, "The god of Islam (Allah) and the god of Christianity (Jehovah), despite their common origin in the god of Judaism (Yahweh), are mutually exclusive. Jehovah and Allah, at least as traditionally understood, cannot both exist at the same time. Both claim to be the Creator of the universe, but they have contradictory attributes (e.g., Christianity claims that there are three 'persons' known as God but Islam claims that there is only one)."[4] There is a great deal of confusion in religious thinking and belief. The religious literature is replete with internal contradictions; the confusion is so chaotic and dense that any effort to resolve it is doomed to fail. For example, one of the attributes of both the Christian God and Allah is that He is incomprehensible. If God indeed is incomprehensible, how can then one describe Him and assign Him any attributes? Should we believe that these attributes (e.g., omnipotent, omniscient, all merciful, etc.) are untrue, or else God, after all, is comprehensible? In spite of His incomprehensibility, so much has been written about God and His attributes, His hell and heaven, His retribution and reward, that no other subject may have received as much attention or written detail.

Comprehending the incomprehensible is indeed a veritable paradox. The picture is much more muddied by the celebrated psychologist, Carl Jung, who had tried to justify the paradoxical description of God and metaphysics pertaining to religion. He said, "Oddly enough, the paradox is one of our most valued spiritual possessions, while uniformity of meaning is a sign of weakness. Hence religion becomes inwardly impoverished when it loses or cuts down its paradoxes; but their multiplication enriches because only the paradox comes anywhere near comprehending the fullness of life. Non-ambiguity and non-contradiction are one-sided and thus unsuited to express the incomprehensible."[3] So, describing God is indeed a conundrum, which does not seem to have a resolution. In view of Jung's way of thinking, any description of God, which is sufficiently muddled and riddled with contradictions, is good and eminently apt. Metaphysicians have rendered religion into a knotted riddle which keeps on becoming more and more intricately tied into itself. The more paradoxical a description of God is, the more beautiful and enchanting it becomes. It is quite another thing that it loses meaning completely, if it had any to start with. A scientist, on the other hand, does not have the same license. For any theory that he propounds or a statement that he makes, he is called upon to provide a proof for it; otherwise his theory and statement do not have any credibility. Jung was fully conscious of this one-sided tilt in favor of religion when he wrote, "The religious minded man is free to accept whatever metaphysical explanation he pleases about the origin of these images; not so the intellect, which must keep strictly to the principles of scientific interpretation and avoid trespassing beyond the bounds of what can be known. Nobody can prevent the believer from accepting God, Purusha, the Atman, or Tao as the Prime Cause and thus putting an end to the fundamental disquiet of man. The scientist is a scrupulous worker; he cannot take heaven by storm. Should he allow himself to be seduced into such an extravagance, he would be sawing off the branch on which he sits."[3] By implication, Jung seems to admit that metaphysics is an extravagance and the metaphysician is an unscrupulous dreamer. He does not have contact with the worldly reality although he may very well believe that he indeed is one with the reality, whatever it is. According to Carnap, "Metaphysicians cannot avoid making their propositions non-verifiable, because if they made them verifiable, the decision about the truth or falsehood of their doctrines would depend on experience and therefore belong to the region of empirical science. This consequence they wish to avoid, because they pretend to teach knowledge which is of a higher level than that of empirical science…and precisely by this procedure they deprive them of any sense."[1]

References

1. Carnap, R. *Philosophy and Logical Syntax*—Chapter on "The Rejection of Metaphysics," 1935, http://nb.vse.cz/-sloukova/FIL.418/carnap.htm.
2. Furst, C. *Origins of the Mind*. Englewood Cliffs, NJ: Prentice Hall, Inc., 1979, p.7.
3. Jung, C. *The Basic Writings of C.G. Jung*—"The Religious and Psychological Problems of Alchemy." Violet Staub De Laszlo, ed. New York: The Modern Library, 1993, pp. 552, 553.
4. Lowder, J. "Is a Proof of the Non-existence of a God Even Possible?" http://www.infidels.org/library/modern/jeff_lowder/ipnagep.htm.
5. Plato. *Five Dialogues*—*Phaedo*. G.M.A. Grube, trans. Indianapolis: Hacket Publishing Company, 1981, p.120.
6. Russell, B. *A History of Western Philosophy*. New York: Simon and Schuster, 1972, pp. 288-89.
7. Smith, G. *Atheism*—*The Case Against God*. Buffalo, NY: Prometheus Books, 1989, p.31.
8. "History of Philosophy of Mind", http://www.Xrefer.com/entry/552820.

(First published at chowk.com on April 12, 2003)

CHAPTER 8

SKEPTICAL AND COUNTER-SKEPTICAL TRENDS IN MEDIEVAL ISLAM

Like his predecessor al-Kindi and his successor Ibn Sina, he (al-Razi) was also a philosopher but unlike them, he made no attempt to reconcile Greek philosophy and Islamic religion. To him the two were irreconcilable. In fact, he was a radical thinker who rejected the concept of prophecy, challenged Koranic dogma, and subordinated theology to philosophy. In this respect, he was rare if not exceptional in Islam (Hitti).[10]

Have you not seen how Allah created the seven heavens one above the other, setting in them the moon as a light and the sun as a lantern? Allah has caused you to grow from the earth, and to it He will return you. Then He will bring you forth. (Quran, ch. 71:15-16)

The tenth principle is that Allah—the Exalted—has sent Prophet Muhammed—the praise and peace be upon him—as the seal of prophets and as an abrogator of all previous religions before him: the religions of the Jews and the Christians and the Sabians. He (Allah) upheld him with unmistakable miracles and wonderful signs such as splitting of the moon, the praise of the pebbles, and causing the dumb animals to speak, as well as water flowing from between his fingers and the unmistakable sign of the Glorious Koran with which he challenged the Arabs (Al-Ghazali).[5]

71

Introduction

The Arabs did not have philosophy, mathematics, or any rational sciences as part of their culture and tradition before the advent of Islam. Although they were aware of Jewish and Christian religions because Jews and Christians lived among them, they themselves were idolaters. They were unsophisticated in their beliefs and outlook. When they had embraced Islam and they conquered territories outside Arabia in the seventh and eighth centuries, they came in contact with other civilizations and cultures, philosophy, and other rational sciences such as mathematics, astronomy, physics, etc., that had been transmitted from Greece into these countries.

During the time of the Abbasi'd Khalifah (Caliph) Mamun-al-Rashid who had established a Bait-el-Hikmah (House of Wisdom) in Baghdad, the influence of the exotic thought seeped into Islamic culture, and its impact on the Arab way of understanding the teachings of Quran was inevitable. Works of Greek philosophy and natural sciences were available in Alexandria, Egypt and some other Syrian cities. Mamun-al-Rashid employed scholars of all religions, Judaism, Christianity, Islam, etc., for the purpose of translating these works into Arabic. In spite of the strong hold of Islamic theological dogma on the minds of the Arabs, skepticism and rational thinking gradually germinated and flowered under the encouragement and protection provided by the Khalifah.

The first skeptics of Islam called themselves Mutazilites, those who keep to themselves. They preached that God was a Perfect Being and took no attributes other than His unity into account. This led to the belief that the text of Quran was created and not eternal (qadim). Encouraged by the spirit of free thought, many notable philosophers and scientists emerged, in due time, who continued with the development of sciences and philosophy building upon the Greek heritage. Al-Kindi was the first such Muslim Arab philosopher who created a doctrine for conciliation between Aristotelian and Platonic philosophies and the Scriptures. This approach became quite popular in later Arabian philosophical thought. He started interpreting Quranic text rationally, and wherever he encountered conflict, he devised an easy escape through allegorical exegesis for the resolution of the conflict. He suggested that wherever the Quranic text appeared contrary to reason, the Quranic text should be interpreted allegorically or symbolically.

Subsequent scholars used this device with mixed results. For example, Ibn Rushd proclaimed, "Since the religion (Islam) is true and summons to the study which leads to knowledge of the Truth, we the Muslims know definitely that demonstrative study does not lead to (conclusions) conflicting with what Scripture has given us; for truth does not oppose truth but accords with it and bears witness to it."[11] Thus, it was a given that the scripture was infallible and true, everything else needed to be brought in harmony with the book by allegorical interpretation, by twisted argument, or simply by subterfuge. Ibn Rushd was later criticized by others for having double standards.

There were hundreds of noteworthy Muslim philosophers, scientists, mathematicians, astronomers, and contributors to other sciences, who had developed original knowledge in the medieval times from which the rest of the world benefited; however, all of them cannot be discussed herein for lack of space and scope. Instead, only a few philosophers who were most prominent in their own time and who have left a permanent imprint on the history of the evolution of sciences and philosophy will be discussed here. Of necessity, the discussion will be restricted to al-Razi, the great doctor of medicine who is latinized in the Western world by the name of Rhazes; Ibn Sina, the successor of al-Razi; the unsurpassed medical doctor and the great philosopher, al-Ghazali; Hujjat-el-Islam (Proof of Islam), a great philosopher who had everlasting influence on Islamic thought and who used rational argument to counter the philosophy of Ibn Sina and al-Farabi; and lastly, Ibn Rushd who is more revered in the Western world than in the Islamic world to which he belonged. Other philosophers will be mentioned en passant only when occasion arises for reference.

Al-Razi—A Believer of God But Not of Prophets

> The smallest measure of original thought, even if it does not reach un-revisable truth, al-Razi insists, helps to free the soul from its thrall in this world and secure for us that immortality which was so wrongly described and so vainly promised by the Prophets (Goodman).[8]

Abu Bakr Muhammad Ibn Zakarya Al-Razi (865-930) was born at Rayy near modern Teheran in 865. He derived his last name al-Razi (or the latinized version Rhazes) from his birthplace. He is mainly remembered for his work in medicine although he was much more versatile and the range of his intellectual activities hardly left any area of knowledge current in his time unexplored and to which he did not make a significant contribution. He was a prolific writer and is said to have written more than two hundred works.

According to Hitti, "Al-Nadim's Fihrist…lists one hundred and thirteen major and twenty-eight minor works by al-Razi. One of his principal works on alchemy, the *Kitab-el-Asrar*, (the book of secrets) was rendered into Latin by the eminent translator Gerard of Cremona and became a chief source of knowledge until superseded in the fourteenth century by Jabir's works."[10] Hitti further dwells on al-Razi's contributions: "Two of his medical works may be singled out: *al-Hawi* (the comprehensive book) and *al-Judari-wal-Hasbah* (small pox and measles). True to its name, *al-Hawi* was a veritable medical encyclopedia summing up what the Arabs knew of Greek, Syriac, Persian, and Hindi medicine and enriched by the addition of the author's experiments and experience."[9]

Al-Hawi was first translated into Latin by the Sicilian Jewish physician Faraj ben Salim. It was printed under a new title *Continents* from 1486 onwards. A fifth edition was printed in Venice in 1542. According to Hitti, "al-Razi's monograph on small pox and measles, an ornament to Arab medical literature, is considered the earliest of its kind. Translated into Latin, it was printed about forty times between 1498 and 1866; it was translated into a number of modern languages including English (1848). It confirmed the author's reputation as one of the keenest thinkers and greatest clinicians not only of Islam but of Christendom."[9]

As recent as May 1970, the World Health Organization (WHO) recognized that "…his (al-Razi's) writings on small pox and measles show originality and accuracy and his essay on infectious diseases was the first treatise on the subject."[4] Writing in their book *The History of Psychiatry*, Alexander and Sheldon noted, "In the field of psychiatry, Rhazes was as good as the finest of the Hippocratic physicians. He was a careful describer of all illnesses, including mental ones. He combined psychological methods and physiological explanations in a way reminiscent of Hippocratics, and he used psycho-therapy in a primitive but dynamic fashion."[2]

Al-Razi's picture hangs in the hall of the Faculty of Medicine in the University of Paris.

While al-Razi's medical contributions are lauded and prominently described by historians, his philosophical work is generally bypassed without much comment. His views about prophet-hood, religion, and divine revelation made him unpopular in the Islamic world. "Given the general repugnance toward al-Razi's philosophical ideas among his contemporaries and medieval successors, few of these works were copied," remarked Goodman.[8] Goodman further stated that his "... other works deal with eros, coitus, nudity and clothing, the fatal effects of Simoom...One work defends the proposition that God does not interfere with the actions of the other agents."[8] Al-Razi described in his *Sira-el-falsafiyya* (Philosophical Way of Life) that "...his has been a life of moderation, excessive only in his devotion to learning, he associated with princes never as a man at arms or an officer of state but always, and only, as a physician and a friend."[8] Al-Razi believed in God but not in prophets. He considered prophets to be impostors.

According to Alexander and Sheldon, "Rhazes always fought charlatanism and stood by his principles as a physician and a man. When the patriarch of Bokhara argued with Rhazes and could not budge the great teacher from his point, he sentenced him to be hit over the head with his own book until the book or the head broke. Rhazes was blinded by this punishment and remained sightless because he would not undergo an operation by a surgeon who was unfamiliar with the anatomy of the eyeball."[2]

Ibn Sina—The Al-Shaykh Al-Rais (The Dean of the Learned)

In the 1950's three millenary (according to the lunar calendar) celebrations were held in Ibn Sina's honor by Persians, Turks, and Arabs. The Turks claimed him because of his father's birthplace Balkh, where Turkish and Persian were spoken. The Teheran and Baghdad celebrations were distinguished by including participants from four continents. A feature of the Teheran celebration was unveiling Ibn Sina's statue at Hamadan, where a new tomb for him was built (Hitti)[9]

Abu Ali al-Hussain Ibn Abdallah Ibn Sina (981-1037), popularly known as Avicenna in the Western world, was born in 981 at Afshana near Bokhara (Transoxania). Although he was Persian (Iranian) by birth, Ibn Sina wrote most of his works in Arabic, the prevalent language of scientific and philosophical expression at that time. Ibn Sina was a truly gifted person and had a knack of absorbing easily whatever he set his eyes on to read.

At the age of ten years, he had memorized Quran, and at the age of sixteen, he had mastery of most of the known sciences at that time. He is said to have had difficulty grasping the Aristotelian metaphysics in the beginning, but when one of al-Farabi's books on the subject came to his hands, it was as if a light shone inside him and his difficulties disappeared. He gained access to the well-stocked library of Bokhara's governor after he cured him of his illness. He virtually devoured the reading material available in the library.

He was born in politically turbulent times and went through ups and downs several times in his life. He wrote his al-Qanun-fi-al-Tibb (popularly known as Canon in the West) at Hamadan. His canon was really encyclopedic in its coverage of the Hippocratic and Galenic traditions "synthesized with Syro-Arabic and Indo-Persian sources and supplemented by the author's experience and experimentation. He made even old material in it look like new and usable: more methodical in arrangement, classification and presentation than al-Hawi."[9] It was translated into Latin by Gerard of Cremona "in the last third of the fifteenth century" (Hitti (9), p.118) and ran into three editions.

There is some confusion here in the chronology. Many believe that the Latin translation of Canon was done by Gerard of Cremona but he couldn't have done it in the fifteenth century because he died in 1187. So it was done in the twelfth century. There are others who believe that Canon was translated by another Gerard in thirteenth century. According to Alexander and Sheldon, "The book (Canon) became the medical bible in Asia and later in Europe and was used until the dawn of anatomical experimentation in the sixteenth century. Robinson, the medical historian, considered the *Canon* the most influential book ever written."[2] Like al-Razi, Ibn Sina wrote prolifically and on every subject that was prevalent in his time. He wrote on mathematics, mechanics, music, philosophy, and even poetry. "His most celebrated Arabic poem describes the descent of soul into the body from the Higher Sphere."[1]

In philosophy, he built on Aristotelian and Neo-Platonic foundations and formulated his own ideology. According to Ibn Sina, "God was the First Cause or Creator, the necessary Being in whom essence and existence were one. From Him there emanated a series of ten intelligences, ranging from the First Intelligence down to the Active Intelligence which governed the world of embodied beings. It was from the Active Intelligence that the ideas were communicated to the human body by a radiation of the divine light, and thus the human soul was created...He (Ibn Sina's God) differs from the Islamic God in the rational interpretation of his attributes and in his creativity. Creation was not ex nihilo. Matter was eternal (qadim) and the process was limited in neither time nor place. It was rather one of his emanations as a consequence of His will and being."[11] This question of the eternity of matter became contentious and was one of the three issues on which al-Ghazali declared al-Farabi and Ibn Sina as kafirs (infidels) and those who believed in it punishable by death.

Ibn Sina also believed that bodily resurrection after the death cannot be explained rationally and is therefore denied. He believed, however, that the soul was eternal and would survive without its physical embodiment. Such verses in Quran that refer to physical resurrection should be understood and explained "allegorically." He also had a different understanding of the nature of prophet-hood than the traditional belief. According to him, "...prophecy was not simply a grace of God; it was a kind of human intellect, and indeed the highest kind. The prophet would participate in the life of the hierarchy of intelligences, and could rise as high as the First Intelligence. This was not an exclusive gift to prophets only, however; the man of high spirited gifts could also attain to it by the way of ascesis."[11]

Ibn Sina was a free thinker who believed in the existence of God; he would not, however, accept facts without analyzing them rationally. His contributions to medical science were monumental, and his impact on medieval Islamic science and philosophy noteworthy. Among many of his views divergent from the Islamic tradition was the issue of free will. "If human beings were controlled by divine necessity, they are not responsible for wrong doing and should not be punished by a just God.

If they are not controlled, God's sovereignty is compromised," reasoned Ibn Sina.[9] According to Quran, "Had Allah pleased, He could have guided all people" (Ch. 13:31).

Ibn Sina's book on philosophy was titled *al-Shifa* (healing) and its abridgment *al-Najah* (deliverance). *Al-Shifa* gained worldwide readership like his *Canon* and was much appreciated for its depth and originality. It was published in the twentieth century in six volumes (Cairo 1952-65) and is considered to be the largest work by one author. Ibn Sina had a kind of free style of living. He rationalized his drinking of wine by suggesting, "By religious law wine is illegal for the fool; by intellectual law it is legal for the intelligent." He said that he started drinking wine to keep awake during night for study. Despite his rationalization, Quran forbids drinking alcohol for everyone.

Ibn Sina was unique and deservingly one of the greatest and most versatile human beings of the last millennium. "An impressive monument to the life and works of the man who became known as the "doctor of doctors" still stands outside Bukhara museum and his portrait hangs in the Hall of the Faculty of Medicine in the University of Paris."[1]

Al-Ghazali—The Greatest Mystical Theologist of Islam

> Our present life in relation to the future is perhaps only a dream, and man, once dead, will see things in direct opposition to those now before his eyes; he will then understand that word of the Quran, "Today we have removed the veil from thine eyes and thy sight is keen."
>
> Such thoughts as these threatened to shake my reason, and I sought to find an escape from them. But how? In order to disentangle the knot of this difficulty, a proof was necessary. Now a proof must be based on primary assumptions, and it was precisely these of which I was in doubt. This unhappy state lasted about two months, during which I was, not, it is true, explicitly or by profession, but morally and essentially, a thoroughgoing skeptic...I owed my deliverance, not to concatenation of proofs and arguments, but to the light which God caused to penetrate into my heart (Al-Ghazali).[6]

Abu Hamid Ibn Muhammad Ibn Muhammad al-Tusi al-Shafi'i al-Ghazali (1058-1111), known as Al-Gazel in the Western world, was born in 1058 in Khorasan, Iran, twenty-one years after the death of Ibn Sina and one hundred and eight years after the death of al-Farabi. He received his education in the prevalent curricula at Nishapur and Baghdad. He excelled in the studies of Islamic theology and philosophy in recognition of which he was appointed a professor of law at the Nizamiyah University of Baghdad that was most prominent at that time. But after a few years, he forsook this prized appointment and became a wandering ascetic.

At the age of thirty-six years, he went through an intellectual crisis, a sort of catharsis, a period of doubt and soul-searching and skepticism. However, he returned to his faith in Islam with renewed vigor and determination and abandoned rationalism that had created the doubts in his mind to start with. He narrates this experience in his book *Munkidh min-al-Dalai* (Confessions, or Deliverance from Error), "Again, the eye sees a star and believes it as large as a piece of gold, but mathematical calculations prove, on the contrary, that it is larger than the earth. These notions, and all others which the senses declare true, are subsequently contradicted of falsity in an irrefragable manner by the verdict of reason. Then I reflected on fundamental principles...Who can guarantee you that you can trust to the evidence of reason more than to that of the senses?"

By tenuous and convoluted arguments against rationalism but wholly satisfactory to himself, al-Ghazali chose the path of divine revelation, in which he had undivided faith, and spent most of his life in waging a vendetta against the philosophical ideas of al-Farabi, Ibn Sina, and others who had adopted Aristotelian and Neo-Platonic philosophy as their point of departure. He wrote his universally renowned book *Tahafut-al-falasifa* (*The Incoherence of Philosophers*) between 1091 and 1095, some fifty-four to fifty-nine years after the death of Ibn Sina. His polemic is mainly aimed at al-Farabi and Ibn Sina.

In this monumental book, al-Ghazali had considered twenty questions attributed to the philosophers, and he refuted them on every count. He used the philosophical argument and approach in pointing out the defects in Ibn Sina's philosophy. The book is known for its excellence of argument and remains a book of reference to this date.

In the conclusion of his *Tahafut*, al-Ghazali stated: "If some one says, 'You have explained the doctrine of these (philosophers); do you then say conclusively that they are infidels and that the killing of those who uphold their beliefs is obligatory?' We say: 'pronouncing them infidels is necessary in three questions.' One of them is the question of the world's pre-eternity and their statement that God's knowledge does not encompass the temporal particulars among individual (existents). The third is their denial of the resurrection of bodies and their assembly at the day of judgment."[7]

Al-Ghazali took issue with the philosophy of al-Farabi and Ibn Sina and conceded, "Regarding mathematical sciences, there is no sense in denying them or disagreeing with them. For these reduce in the final analysis to arithmetic and geometry."[7] The verdict dealt by al-Ghazali against philosophy proved fatal for further development of analytical thought in Islamic society. Although he seems to have tolerated the study of mathematics as seen above, in fact, five years after writing his *Tahafut*, he very much discouraged the study of mathematics, also, as we shall see herein later. The study of philosophy in particular and of rational sciences in general was not encouraged in the Islamic world later on, and in due time, these sciences became foreign to the Muslim culture. Even in modern times, rational sciences and philosophy do not have deep and self-sustaining roots in Muslim society; they are imported and planted into the society from abroad. Al-Ghazali classifies the philosophical system into three sub-systems, in his Munkidh, as follows.

1. The Materialists: They reject an intelligent and Omnipotent Creator and disposer of the universe. In their view, the world exists from all eternity and had no author. The animal comes from semen and semen from the animal; so it had always been and will always be; those who maintain this doctrine are atheists.

2. The Naturalists: They devote themselves to the study of nature and the marvelous phenomenon of the animal and vegetable world...Acknowledging neither a recompense for good deeds nor a punishment for evil ones, they fling off all authority and

plunge into sensual pleasures with the avidity of brutes. These also ought to be called atheists.

3. The Theists: This school refuted the systems of the two others, i.e., the Materialists and Naturalists; but in exposing their mistakes and perverse beliefs, they made use of arguments, which they should not. "God suffices to protect the faithful in war" (Quran, 33:25).[6]

Al-Ghazali divided the philosophic sciences into six categories, the first of which is mathematics. His attack on mathematics is guarded but sure. He asserted in his Munkidh, "Falling a prey to their passions, to a besotted vanity, and the wish to pass for learned men, they persist in maintaining the pre-eminence of mathematicians in all branches of knowledge. This is a serious evil, and for this reason those who study mathematics should be checked from going too far in their researches... It is rarely that a man devotes himself to it without robbing himself of his faith and casting off the restraints of religion."[6]

The philosophical concepts change with time; mathematical and scientific theories grow and evolve with time. Divergent views are not worthless; in fact, they help to advance knowledge. Banning the study of natural sciences out of fear that people may lose faith is ludicrous and a step in the wrong direction. It is only reasonable that difference of cognition should be recognized and accommodated. It is preposterous and inhuman to declare those who do not testify to the religious dogma punishable by death. Freedom of individual thought and choice should be an inalienable right of people. Expulsion of rationalism from man's intellectual engagements in order to protect religious dogma is the worst kind of discrimination.

If a religion cannot stand the test of reason and common sense, why defend it? Religion should be an individual's own choice and should not be imposed by the clergy. The notion that reason is not complete and comprehensive may be true, but if it is allowed to mature, it grows and keeps on refining itself. Self-correction is the best thing by which the rational sciences refine themselves; religion, on the other hand, is petrified and rigid. It is bound to become outdated in due time.

The scientific edifice is built brick by brick; scientific truths are not attained by a leap of faith. A leap of faith is not very certain. Reason is not perfect, but what good is there in replacing reason by religion whose mainstay is blind faith? What guarantee is there that the blind faith is truer and better than a rational truth? A society that banishes reason and rational sciences from its culture becomes a barren society and is doomed to degrade and decay in spite of the supremely divine and unblemished religion it may claim to possess.

Islamic society is a classic example of this degradation and decay. During the times of Al-Razi and Ibn Sina, the Muslim society was the apex of excellence in terms of scientific achievements. After Al-Ghazali banished the philosophical and scientific pursuits from the Muslim society, it gradually and persistently sank to its present decadent condition. In place of rational thought, look at what Al-Ghazali offered: "If the child were to be brought up on this firm belief then occupy him(self) with gaining his livelihood, he might not be more enlightened. But according to the belief of the people of the truth he will be saved. This is because the religion did not obligate the uncivilized Arabs more than believing and certifying in the apparent articles of belief. They were never obligated to research, inquire, nor to be burdened with the classification of arguments."[5] At another place, he offers, "To teach them disputation is decidedly harmful to them as it will perhaps arouse doubts in their minds which will shake their belief. Once these doubts are aroused, it will not be possible to remedy their shaken belief."[5]

Al-Ghazali also prescribed that the faith in the "Bridge which is stretched over Hell and is finer than a hair and sharper than the edge of the sword" is obligatory.[5] It should be understood literally and not metaphorically because "Allah who is able to make the birds fly in the air is also able to make mankind walk over the bridge." Al-Ghazali's influence against nurturing free thought that leads to the developments of material sciences proved a mortal blow to Muslim society from which it has still not recovered. Muslim society is confused and perplexed. It wants the fruits of the material sciences and technological advancements, without which it is difficult to survive with dignity, yet it does not want to loosen the grip of religion as codified by Al-Ghazali and his successors. Believing literally in the "allegorical parable" of the bridge stretched over hell is nothing but anachronism in the age of space exploration and nuclear research. If religious faith one must have, let the allegorical device be used to make it appropriate for modern times.

Sufficient space should be allowed to those people who believe, for their own reasons, that religion has become redundant and has outlived its usefulness to human society. One should be tolerant of divergent views. There are hundreds of religions in the world, each one of them claiming to be the only true religion. All of them cannot be right; on the other hand, there is a greater probability of all of them being wrong.

Ibn Rushd—The Last Great Philohopher of Islamic Spain

> Averroes was a confirmed Aristotelian but compromised with religion by maintaining that there is a "double truth," one begotten by faith and the other from "rational philosophy.' This compromise was important for medical psychology: it established the tradition of a medical man keeping his religious convictions and still believing in scientific discoveries (Alexander and Sheldon) 2

Abu-al-Walid Muhammad Ibn-Ahmad Ibn-Muhammad Ibn-Rushd (1128-1199), popularly known as Averroes in the West, was born in 1128, seventeen years after Al-Ghazali's death, in Cordoba into a family of juriscorsults. His father and grandfather were Grand Qadi's (chief judges) in Cordoba at a time when Muslim Spain was a leading center of unparalleled excellence in every respect in Europe. Ibn Rushd received his education in the curricula that were current in those days at the mosque-based university of Cordoba and specialized in law and medicine.

Philosophy was included in the curriculum of medicine under one title, hikmah. At the invitation of Abu-Yaqub Yusuf, the khalifah of Morocco, Ibn Rushd went over to Morocco. He was commissioned to prepare a "simplified and meaningful text on philosophy. The commission carried with it an honorarium, a robe of honor, and appointment as Chief Justice (Grand Qadi), first in Seville and then in Cordoba."[9] Ibn Rushd was appointed as Khalifah's personal physician in 1182. After Abu-Yaqub's death, his son, Yaqub-al-Mansur, became the khalifah.

According to Hitti, "The cordial relationship between the new patron and his protégé was interrupted in 1194 when the king sent the sixty-eight-year-old scholar into exile and ordered burning of his books. The reason for the unexplained action is not difficult to unearth. Theologians exerted pressure to have the philosopher considered a traitor to his religion…Ibn Rushd's attempt to keep one foot in Islam and the other into the realm of philosophy was no more successful than Ibn Sina's or Al-Kindi's. Two years later the expatriate was reinstated in royal favor but it was too late. Humiliated and heart-broken over the destruction of his books, the aged philosopher died on December 10, 1198. His remains were later removed to Cordoba."[9]

Although Ibn Rushd is better known for his commentaries on Aristotle's philosophy, his contributions in medicine and other fields were also remarkable. His book on medicine entitled *al-Kulliyat-fi-al-Tibb*, latinized as *Colliget*, was translated into Hebrew and Latin and was used as a textbook in Europe, but it was no match for Ibn Sina's *al-Qanun* or Al-Razi's *Al-Hawi*. He wrote a treatise on the motion of the planets also which was entitled *Kitab-fi-Harkat-al-Falak* (Book on the Heavenly Movements). He contributed to jurisprudence (Fiqh). He belonged to Maliki School of Fiqh. He wrote thirty-eight commentaries on Aristotle's philosophy, of which thirty-six survived in Hebrew, thirty-four in Latin, and only twenty-four in Arabic. According to Hitti, "Within fifty years after his death, Averroes, to use his Latin name (as it came from Hebrew), became known as 'the great commentator,' and Averroism, his brand of philosophy, achieved currency."

His epochal book, however, is *Tahafut Al-Tahafut* (The Incoherence of the Incoherence). In this book, Ibn Rushd discussed Al-Ghazali's twenty questions on which he (Al-Ghazali) criticized and refuted the philosophy of Al-Farabi and Ibn Sina and attempted to show defects in Al-Ghazali's arguments. His attempt was noteworthy but failed to soften the deadly impact of Al-Ghazali's verdict on the status of philosophy in the Islamic world. Ibn Rushd tried to build a bridge between philosophy and the divinely revealed knowledge but did not succeed, in as much as the Islamic world was concerned.

Some of his rationalizations appear farfetched; Al-Razi's statement that rationalism and revelation cannot be reconciled resurfaces in this perspective. For instance, regarding bodily resurrection after death, Ibn Rushd goes into a lengthy discussion, partially agreeing with Al-Ghazali, and suggests, "…it must be assumed that what arises from the dead is simulacra of these earthly bodies, not these bodies themselves, for that which has perished does not return individually and a thing can only return as an image of that which has perished, not as being identical with what has perished. Therefore the doctrine of resurrection of those theologians who believe that the soul is accident and that the bodies which arise are identical with those that perished can not be true."[12] The verses on resurrection appear in several chapters in the Quran. For instance: "They say what! When we are reduced to bones and dust, should we really be raised up (to be) a new creation' Say: (Nay!) be ye stones or iron, or created matter which, in your minds, is hardest (to be raised up), (yet shall ye be raised up)!" (Quran, ch.17:49-51).

Similarly, there is disagreement between Al-Ghazali and the philosophers regarding the miracles. The miracle of Moses' staff turning into a serpent that devoured the snakes of the magicians as described in Quran is interpreted by the philosophers as "the refutation by the divine proof, manifest at the hand of Moses, of the doubts of those who deny (the one God)."[7] The philosophers denied the splitting of the moon because they "claim that there has been no soundly transmitted, indubitable reporting of it." Ibn Rushd is somewhat evasive and ambiguous on this matter when he states: "Most things which are possible in themselves are impossible for man, and what is true of the prophet, that he can interrupt the ordinary course of nature, is impossible for man, but possible in itself; and because of this one need not assume that things logically impossible are possible for the prophets, and if you observe those miracles whose existence is confirmed, you will find that they are of this kind."[12]

In spite of his guarded caution in denying the miracles outright, Ibn Rushd could not save himself from the wrath of the fanatics and blind believers of his time who set his books on fire because they believed him to be an atheist. Both Al-Ghazali and Ibn Rushd were excellent scholars, the former steeped to his skin in theology and had absolute faith in the scriptures and the latter having an open, though not completely, mind and who was a defender of reason and rationalism, yet they were so very different in interpreting the same basic issue and arrived at completely different results. The former is celebrated in the Islamic world and the other is largely ignored but has been accorded a unique position in the history of philosophic thought and science, in the West. According to Karen Armstrong, "Ibn Rushd was a revered but secondary figure in Islam, but he became very important indeed in the West, which discovered Aristotle through him."

An interesting question comes to mind: Were both of them (Al-Ghazali and Ibn Rushd) living in the modern time, what kind of scholars would they be? Al-Ghazali, most probably, would still be Al-Ghazali, an influential Ayatollah beckoning people to the purity of his brand of Islam. Ibn Rushd, on the other hand, would perhaps be a much more liberated philosopher in his outlook with his analytical skills honed by the knowledge of the modern empirical sciences and would be much more forthright in expressing his ideas freely and fearlessly. He would probably not need the doctrine of "double truth" to uphold that which cannot be upheld. He would be more like Al-Razi. For that matter, Al-Razi was a modernist who never cared to concoct arguments to rationalize religious belief to make it more akin to reason. He considered such efforts a mere waste of time and an exercise in intellectual dishonesty.

References

1. Ahmad, M.. "Ibn Sina (Avicenna)—Doctor of Doctors," http://www.ummah.net/history/scholars/ibn sina/

2. Alexander. F.G., and T.S. Sheldon. *The History of Psychiatry*. New York Harper and Row Publishers, 1966, pp.64, 62-63.

3. Armstrong, K. *A History of God*. New York: Ballantine Books, 1993, p.194.

4. "An Introduction to Al-Razi," http://library.thinkquest. org/17137/Main/Math Science/Medicine/al razi.html

5. Al-Ghazali. *The Foundations of the Islamic Belief.* Shaykh Ahmad, trans. Darwish Mosque of the Internat, P.O. Box 601, Tesque, NM 87574,USA, Chapter 3, Pillar 3, p.21 of 26.

6. Al-Ghazali. *"Munkidh min al-Dalal"* (Confessions, or Deliverance from Error, Fordham.edu/halsal/1100 ghazali-truth.html

7. Al-Ghazali. *Tahafut-al-falasifa* (The Incoherence of the Philosophers). M.E. Marmura, trans. Utah: Brigham Young University Press, 1997, pp. 230, 166, 11.

8. Goodman. L.E. "al-Razi." *The Encyclopedia of Islam*. C.E. Bosworth et al, eds., 1995, pp. 474-77.

9. Hitti, P.K. *Islam—A Way of Life*. Chicago: Henry Regnery Company, 1970.

10. Hitti, P.K. *History of the Arabs*. New York: St. Martin's Press, p. 435.

11. Hourani, A. *A History of the Arab People*. Warner books, A Time-Warner Company, 1991, pp. 174-75.

12. Ibn Rushd. *Tuhafut Al-Tahafut* (*The Incoherence of The Incoherence*). Simon Van Den Bergh, trans. Published by The Trustees of the E.J.W. Gibb Memorial, 1987, pp. 362, 315.

(First published at chowk.com on November 28, 2002)

CHAPTER 9

CONFLICT OF SCIENCE
WITH THEOCRACY

The question of whether science can survive in a theocratic environment had been gnawing at my mind for a long time. This question had arisen from another one: Why has scientific research not flourished in the Islamic world? Although all the various governments in the Muslim countries are not constitutionally theocratic in structure, all of them do indeed have constitutional provisions forbidding divergence from the fundamental beliefs based on religious tradition and the holy scriptures. Any deviation is punishable severely. In controversial situations, the benefit of the doubt invariably goes to the religious tradition.

The more I pondered over this question, the more convinced I became that the two, i.e. science and theocracy, do not fit with each other. If one is a circular hole, the other is square peg. The reason that the two cannot coexist amicably is that they are fundamentally different from each other in concept and practice. The cornerstone of theocracy is belief in the existence of supernatural God(s) who has designed the whole universe according to a preconceived intelligent plan and has preordained what the humans have to believe uncritically and how they have to live their lives. On the other hand, science is founded on the precept that there are certain fundamental laws of nature according to which the universe is operating. These laws are universal and do not change with time, or at least they have not changed in as much as is known to us. There is no evidence that God ever fiddled with them to suit His whims. Science continually keeps on verifying its laws, hypotheses, and concepts according to the empirical evidence that becomes available from observations and measurements. Thus science keeps on correcting itself.

The thesis that "it happens like that because it's the will of God" is not scientific and hence it's not credible in science. It's quite acceptable however from a religious viewpoint. The foundations of religious belief are immutable; there is no change or variation in them. The "word of God" is eternal and not susceptible to change. Science does not believe in such presumptions.

The first head-on clash between science and theocracy, which is well documented in the history of science, was between Galileo and the Holy Inquisition which had been set up by the Roman Catholic Church in the seventeenth century. This incident has been reported so frequently that even those who do not care much for its details are aware of it. A little repetition will do no harm and in fact is quite appropriate in the present context.

According to Ptolemaic conception of the movements of the planets, Earth was believed to be the center of the universe and stationary. This concept somehow found its way into the Holy Bible and became an article of Christian faith. The Ptolemaic concept is called the "geocentric theory." In due time, it was determined that the geocentric theory was unwieldy and not amenable for easy and accurate astronomical computations. The "heliocentric" theory, which had the Sun at the center of the solar system, yielded better results. With this in view, Copernicus formally propounded the heliocentric theory in his book *De revolutionibus orbium coelestium* but did not publish it until late in his life for fear of retribution from the Roman Church. He is said "... to have received a copy of the printed book for the first time on his deathbed."[3].

His theory became widely known to researchers after the publication of the book. Many renowned astronomers including Galileo and Kepler accepted the heliocentric theory but Tacho Brahe, another notable astronomer, did not accept that Earth was not stationary. Copernicus's theory was formally validated by Newton's theory of gravitation about 150 years later.

Galileo supported heliocentric theory and was confronted by the Holy Inquisition, which compelled him to recant his heliocentric thesis. See the text of his abjuration in n.1. He was confined to house arrest in which he finally died. The church rescinded its position in 1992 (n.2), after more than 300 years and recognized the heliocentric system as valid.

Galileo had other issues also with the Roman Church besides the revolving Earth. It was also believed by the clergy that God had created the heavenly bodies without any flaws. Galileo, using his telescope, found spots on the Sun and mounds and craters on the moon's surface. Galileo used to persuade people to peek through his telescope to see those flaws for themselves. Many of them refused to do so, and many others believed that the inherent reality of the material objects changed when viewed through the telescope.

There was no serious conflict again of the same gravity as that of Galileo in the Christian world. Science and religion had quietly resolved to come to terms and live in peace although there were still arguments and heated debates on issues on which science conflicted with religion.

For instance, when Darwin propounded his theory of evolution by natural selection in the 1850s, the clergy rose in rage against it, piling all kinds of ridicule and vituperation on the evolutionists. The tug-of-war is still continuing but with the passage of time, science is gaining ground on the basis of empirical evidence that has been gathered by the scientists. Religion is again on the retreat and losing ground rapidly. The creation story told in Genesis in the Holy Bible is not going to survive and will need to be revised. Whether the Christian faith and other faiths, which believe in the creation of the universe by an act of God (Be, and it is), can recover remains to be seen. The conflict is so grievous and fundamental that Arthur Peacocke, who is a priest and Canon in the Church of England and until recently was director of the Ramsay Center for the Study of Science and Religion at Oxford University (also winner of the Templeton Prize for Progress in Religion), found it hopelessly difficult to defend the Biblical Genesis and has formulated a new "Genesis for the third millennium" as follows:

There was God, And God was All That Was. God's Love overflowed and God said, 'Let Other be. And let it have the capacity to become what it might be, making it make itself—and let it explore its potentialities.'

And there was Other in God, a field of energy, vibrating energy—but no matter, space, time or form. Obeying its given laws and with one intensely hot surge of energy—a hot big bang—this Other exploded as the Universe from a point twelve or so billion years ago in our time, thereby making space.

Vibrating fundamental particles appeared, expanded and expanded, and cooled into clouds of gas, bathed into swirling whirlpools of matter and light—a billion galaxies.

Five billion years ago, one star in one galaxy—our Sun—became surrounded by matter as planets. One of them was our Earth. On Earth, the assembly of atoms and the temperature became just right to allow water and solid rock to form. Continents and mountains grew and in some deep wet crevice, or pool, or deep in the sea, just over three billion years ago some molecules became large and complex enough to make copies of themselves and became the first specks of life.

Life multiplied in the seas, diversifying and becoming more and more complex. Five hundred million years ago, creatures with solid skeletons—the vertebrates—appeared. Algae in the sea and green plants on land changed the atmosphere by making oxygen. Then three hundred million years ago, certain fish learned to crawl from the sea and live on the edge of land, breathing that oxygen from the air. Now life burst into many forms—reptiles, mammals (and dinosaurs) on land—reptiles and birds in the air. Over millions of years the mammals developed complex brains that enabled them to learn. Among these were creatures who lived in trees. From these our first ancestors derived and then, only forty thousand years ago, the first men and women appeared. They began to know about themselves and what they were doing—they were not only conscious but also self-conscious. The first word, the first laugh were heard. The first paintings were made. The first sense of a destiny beyond—with the first signs of hope, for these people buried their dead with ritual. The first prayers were made to the One who made All-That Is and All-That-Is-Becoming—the first experiences of goodness, beauty and truth—but also of their opposites, for human beings were free.

Apart from the inclusion of God and the One at the outset, the new Genesis is indeed an epitome of the theory of evolution. It seems only a matter of time when the theory of evolution, like heliocentric theory, will find universal recognition. Many theologians of Christian faith with scientific back-grounds, like Reverend Canon Peacocke, are already seeking accommodation with the theory of evolution despite the ongoing vendetta waged by the creationists, as evidenced by Peacocke's formulation. Whether the theory is vindicated in every detail or not, Biblical Genesis, according to which God created the universe in six days and He rested on the seventh, is not going to hold water. Earth is also not as recent as the creationists tenaciously believe; the majority of Christians already believe Earth to be several billion years old.

However, the situation is more serious in the extremely conservative Muslim world. In the face of orthodox beliefs and continued insistence on rejecting everything that does not seem to conform to the traditional beliefs, it is easy to understand why a Muslim scientist did not propound a theory of evolution. The very idea is repugnant to traditional belief and the propounder would surely have been killed under the "blasphemy law." This also is the reason that no substantive contribution was made to science by any Muslim scientist for the last seven to eight hundred years. The Muslim world has factually isolated itself from the mainstream scientific community.

Consider Dr. Aghajari's (a PhD in History) plight in Iran. He was sentenced to death in Iran last year (2002) for criticizing the practice of taqlid (blind imitation). The actual charge filed against him was "insulting and weakening clerics in the name of intellectualism and reform."[1] In Pakistan, people have been victimized and properties vandalized on such a trivial charge as disrespecting beard. Al-Ghazali had condemned to death, Al-Farabi and Ibn Sina posthumously, and all other philosophers who dared to believe that dead bodies would not be physically resurrected in their original earthly forms on the Day of Judgment.[2] He barely tolerated the study of mathematics because he thought pursuit of such knowledge would lead the faithful astray. Symbiosis of science and theocracy is impossible under such extreme conditions. Religion needs to accommodate (or else become extinct in due time) scientific research, free thought, and pursuit of liberal arts (philosophy, etc.,) if the Muslim world is to develop materially and hold its own against other developed nations of the world.

Notes

n.1. Galileo's Abjuration

Following is the text of Galileo's abjuration in front of the Holy Tribunal. The text was presented to Galileo by the Holy Tribunal and he read it. Dressed in the white robe of the penitent, the accused then knelt and abjured as ordered:

I, Galileo, son of the late Vincenzio Galiei, Florentine, aged 70 years, arraigned personally before this tribunal, and kneeling before You, Most Eminent and Reverend Lord Cardinals, Inqisitors-General against heretical depravity throughout the Christian commonwealth, having before my eyes and touching with my hands the Holy Gospels, swear that I have always believed, I believe now, and with God's help I will in future believe all that is held, preached, and taught by the Holy Catholic and Apostolic Church. But whereas—after having been admonished by this Holy Office entirely to abandon the false opinion that the Sun is the center of the world and immovable, and the Earth is not the center of the same and that it moves, and that I must not hold, defend, nor teach in any manner whatever, either orally or in writing, the said false doctrine, and after it had been notified to me that the said doctrine was contrary to Holy Writ—I wrote and caused to be printed a book in which I treat of the already condemned doctrine, and adduced arguments of much efficacy in its favor, without arriving at any solution: I have been judged vehemently suspected of heresy, that is of having held and believed that the Sun is the center of the world and immovable, and that the Earth is not the center and moves.

Therefore, wishing to remove from the minds of your Eminences and all faithful Christians this vehement suspicion justly conceived against me, I abjure with a sincere heart and unfeigned faith, I curse and detest the said errors and heresies, and generally all and every error and sect contrary to the Holy Catholic Church. And I swear that for the future I will never again say nor assert in speaking or writing such things as may bring upon me similar suspicion: and if I know any heretic, or person suspected of heresy, I will denounce him to this Holy Office, or to the Inquisitor or Ordinary of the place where I may be. I also swear and promise to adopt and observe entirely all the penances which have been or may be imposed on me by this Holy Office. And if I contravene any of these said promises, protests, or oaths (which God forbid!), I submit myself to all the pains and penalties imposed and promulgated by the sacred Canons and other Decrees, general and particular, against such offenders. So help me God and these His Holy Gospels, which I touch with my own hands.

I, the said Galileo Galilei, have abjured, sworn, promised, and bound myself as above; and in witness of the truth, with my own hand have subscribed the present document of my abjuration, and have recited it word by word in Rome, at the Convent of the Minerva, this 22nd day of June 1633.

I, Galileo Galilei, have abjured as above, with my own hand.[5]

n.2. Pope John Paul II's Statement

According to Sobel, "Pope John Paul II (in 1992) publicly endorses Galileo's philosophy, noting how 'intelligibility, attested to by the marvelous discoveries of science and technology, leads us, in the last analysis, to that transcendent and primordial thought imprinted on all things."[5]

References

1. Gill, Mohammad. "Religious Conservatism and Modernity Again on Collision Course in Iran," www.pakistanlink.com/Opinion/2002/Nov/29/05.html
2. Gill, Mohammad. "Skeptical and Counter-Skeptical Trends in Medieval Islam," www.chowk.com, November 28, 2002.
3. O'Connor, J.J. "Nicolaus Copernicus," http://www.gap.dcs.st-and. ac.uk/~history/Mathematicians/Copernicus.html
4. Peacocke, A. *Paths from Science Towards God, One World.* Oxford, 2001, pp. 1,2.
5. Sobel, D. *Galileo's Daughter.* New York: Walker & Company, 1999, pp. 275-77, 374.

(First published at chowk.com on September 1, 2005)

CHAPTER 10

DEEP ROOTS OF
RELIGIOUS ORTHODOXY

Religions seldom change, or change much, due to the conscious efforts of the human mortals. The reason for this immutability is that a concept of divinity is built into the structure of almost every religion, and divinity is unchangeable by definition. Yet religions do change with the passage of time; such changes may not be substantial but still they are significant enough to undermine the religious foundation even ever so slightly.

As an example, the Bible considered Earth as an immobile center of the universe before Copernicus. Copernicus's cosmology moved the Earth from the center of the universe (there is no center of the universe) and made it an ordinary planet orbiting around the Sun which originally was believed to be orbiting the Earth. Galileo was compelled by the Roman Catholic Church to recant from his Copernican position regarding the status of the Earth, but in 1992, after more than three hundred years of muzzling Galileo, the Roman Catholic Church renounced its wrong position formally to accept Galileo's cosmic visualization as correct. How the Pope reconciled this fact with the biblical understanding is somewhat of a mystery. The Bible at more than one place stated implicitly or explicitly that the Sun moved around Earth. This language still exists in the Bible which is believed to be the eternal "word of God" by the Christian orthodoxy. There are some other instances of similar discrepancies in the Bible. Similar instances do exist in the Holy Quran also although the Muslim scholars do not dare to discuss them in the open for fear of the stigma of blasphemy.

Such changes as take place in the religions occur due to social and scientific developments with the passage of time. The church resisted the change in its geocentric belief due to Galileo's efforts, but the Christian religion accepted Galileo's view nonetheless, and the Pope formalized its recognition after more than three hundred years of poignant silence. This change in the outlook did not alter the biblical text, which remains as it was before Galileo and Copernicus.

There are numerous skeptical philosophers and scientists in the West who have questioned the foundations of the Bible and lost faith in it. More importantly, there are several people of the clergy who have underscored the need for reinterpreting the outdated scriptures in order to make them more amenable and harmonious with the modern times. One such prominent scholar is Bishop (retired) John Shelby Spong. I shall discuss herein the deep roots of orthodox belief in religion quoting examples of Christianity derived from Spong's writings.

Reinterpretation doesn't change the literal words of the scriptures; reinterpretation is provided in the books of commentaries (Tafsirs, for instance, of the Holy Quran). The commentaries can change in time and be made commensurate with the temporal needs. This is a difficult and challenging task but worth undertaking.

Another motivation for the reinterpretation is that the scriptural language is ancient and archaic. Modern Arabic, for instance, is much different from the Arabic language used in the Quran. A given word in the Quran may have a different meaning these days than what it had fourteen years ago. The situation is more complex in the case of the Bible, which has passed translation from Hebrew to Greek to Latin and to English. Some essence gets lost in translating from one language into another through a concatenation of other intermediate languages.

The greatest motivation for reinterpreting the scriptures is due to the great developments that have taken place in modern times. The laws and codes that had been written in the past for simple life in the desert of Arabia, for example, need to be suitably adjusted for use in different climes and times. The literal meaning needs to be modified retaining the spirit of the injunction. Even God wouldn't like that the same old laws were being used for the greatly different living conditions of modern times.

The orthodox believers, however, generally resist any reinterpretation of the scriptures; for them, the literal comprehension, however discordant with modern life, is sufficient and meaningful in some strange way. Their perpetual slogan is "go back to the fundamentals," which means literal interpretation of the scriptures.

Bishop Spong and the Inconsistencies of the Bible

John Shelby Spong retired in 2000 from his office as the bishop of the Diocese of Newark. He has very good academic, literary, and professional credentials. He earned his first degree (A.B.) in philosophy with a minor in zoology. In honor of his impressive scholarship, he was awarded the degree of Doctor of Divinity by St. Paul's College in 1976 and Virginia Theological Seminary in 1977. He was also honored with Doctor of Humane Letters by Muhlenberg College in 1998. In due time, he acquired a rationalistic attitude and examined the scriptures rather critically to understand if they could be meaningfully used in modern times.

He came to rather severely skeptical conclusions regarding the originality and the traditional orthodox comprehension of the Bible. According to him, "...everything written in the Bible is first of all not eternal, and, second, not necessarily true."[1] In his book *Living in Sin*, he asserted, "In the Bible, there are conflicting accounts of creation, conflicting versions of the Ten Commandments, conflicting understanding of who Jesus is and was...Despite the fact that these conflicts and alternatives are present in Scripture, there are still some who insist that the Bible is inerrant and that its text can be quoted to define and support a wide variety of moral activities."[2] In view of these observations, he declared, "The Bible becomes not a literal road map to reality, but a historical narrative of the journey our religious forbears made in the eternal human quest to understand life, the world, themselves, and God."[3]

In his book *Rescuing the Bible from Fundamentalism*, he wrote, "...there are concepts in the Bible that are repugnant to the modern consciousness. There is a vicious tribal code of ethics that prohibits internally behavior that is actually encouraged in dealing with outsiders.

Moses was a murderer, but this was not a character flaw because his victim was an Egyptian. (Exodus 2:11 ff)...Captive peoples, if spared from death, were reduced to slavery. Captive women were used for sexual sport by their Hebrew conquerors."[4]

In his book *A New Christianity for a New World*, he wrote, "This God is described in our scriptures as hammering the Egyptians with plague after plague, for example, one of which involved the murder of the firstborn male of every Egyptian household in a divine campaign to free the chosen people from slavery (Exodus 7-10). Then this God opened the Red Sea to allow the Hebrews to escape their life of bondage and closed the Red Sea just in time to drown the pursuing army of the Egyptians (Exodus 14). Is that the handiwork of a moral deity?" He continued, "The theistic God of scripture is also said to have stopped the sun in the sky (as if the sun actually rotated around the earth) to enable Joshua to have sufficient daylight to slaughter the Amorites in a battle (Joshua 10). Is that a justifiable cause for divine action?"[5]

In view of the obvious incredibility of these stories, Spong who has a modern and rationalistic outlook finds it impossible to believe in them. The traditional orthodoxy, however, still believes in such stories in spite of the fact that they point to the irrational partiality of God and contradict the modern discoveries of astronomy and cosmology. A God who is defined by such theism, Spong calls a theistic God.

Death of Theistic God

> When people say today, for example, that the "age of miracles is over" what they mean is not that miracles no longer occur, but that they never did—the age when we perceived events as miraculous is gone (John Shelby Spong).

Spong defines a theistic God as "a being, supernatural in power, dwelling outside this world and invading the world periodically to accomplish the divine will."[5] He is the God of rewards and retribution. He will resurrect all the dead bodies on the Day of Judgment to determine who should be rewarded with entry to paradise, an abode of eternal bliss, for his (her) good actions, and who should be condemned eternally to burn in hell for ill deeds. Such a theistic God is dying, Spong proclaims, if not already dead.

What I intend to do here is to provide Bishop Spong's rationale for updating the intent and meaning of the Bible for the modern man. Similar need exists in the Muslim world, also, where the situation is much more complex. Many orthodox Muslims believe that the door of ijtehad (reinterpretation) is shut after Imam al-Ghazali, Hujjat-el-Islam, the Revivalist of the first millennium. Many others believe that Sheikh Ahmad Sirhindi, Mujaddad Aleph Thani (Revivalist of the second millennium) was the last Mujtahed. Yet many others recognize the need for reinterpretation, but they prescribe such impossibly narrow qualifications for such a mujtahed that it is nearly impossible to find such a living person.

Be that as it may, I will try to demonstrate here using generous quotations from Spong, how archaic and anachronistic have the scriptures become in their literal comprehension. He wrote, "The God of theism is so visibly dying that only by playing a game of denial and illusion—a game that many play—can we continue to maintain that this God is still real. That is the nature of the religious dilemma of our generation."[6]

In case anybody misunderstood Spong's proclamation, he wrote, "God, understood theistically, is thus quite clearly a human construct. Please let that fact register. The theistic definition of God is a human creation. Thus theism is not the same as God. Neither is it any more eternal than any other human definition."[7]

In spite of such vitriolic criticism of traditional Christianity and theism, Spong is a self-proclaimed believer. He opened his first chapter "A Place to Begin" in *A New Christianity for a New World* with, "I am a Christian. For forty-five years, I have served the Christian church as a deacon, priest, and bishop. I continue to serve that church today in a wide variety of ways in my official retirement. I believe that God is real and that I live deeply and significantly as one related to that divine reality." He further elaborated, "I have tried for a lifetime to live faithfully, if not comfortably, within the confining boundaries of this institution called the church. That church has conferred on me gifts of honor, position, leadership. and influence. I have loved my life as one of the church's ordained servants. I have never wished the church harm. But I no longer believe that this institution—or the Christian faith as this church has traditionally proclaimed it—can continue to live without dramatic change in our post-theistic world. Somewhere along the way, we Christians appear to have lost the ability to initiate effective self-reformation. We have warned others against idolatry but have not listened to our own warning. We have acted time after time as if the God we have experienced could be or has been captured in and bound by the words of our scriptures, our creeds, and our doctrines. Throughout Christian history, we have acted as if God had to be protected and defended by those in possession of the infinite truth of the divine being. We have presumed that the doorway to God is in our hands. Proclaiming that God can be approached only through our symbols, we have excluded those human beings who do not use our words or refuse to bow before our altars erected by the community of 'the saved.'"

Bishop Spong's self-scrutiny of his faith is a good omen for reformation in the religious institutions, in general. If his example is followed by the scholars and ulema of other religions, there is hope that religion can still survive as a spiritual institution. The religious orthodoxy will then cease to exist and be replaced by religious modernity. This is probably the only way in which religion can become less oppressive and gain a spell of meaningful existence.

Other spades that Spong has stoked in the religious fires are numerous such as the issue of equality of genders; homosexuality, which he considers as a biological condition (like left handedness, for example) and not a human choice and preference; negation of Christ's resurrection and ascension to heaven; the Flood and Noah's Ark; and original sin, among many others, and cannot be adequately discussed here. The intent here is to open up meaningful discussion on the viability of traditional orthodoxy.

Conclusion

The religious orthodox views are ubiquitous; they exist in more or less every religion. Their incompatibility with the evolving scientific worldview is becoming increasingly transparent. Even the entrenched orthodox religionists have relinquished their belief that the Earth is stationary; many others have also abandoned their belief that the Earth was created six thousand years ago.

The creationists are fighting a losing battle against the biological evolutionists. The traditional theistic worldview is outdated and doomed; people will have to let it go in due time, not in the distant future but in a matter of few decades.

At the same time, spirituality is almost a human instinct. Science cannot be a substitute for religion because religion satisfies the spiritual needs of humankind. The majority of people will always believe in one form or another of religion, to wit: John Shelby Spong and his undying faith in Christianity in spite of his very dim view of traditional Christianity.

References

1. Spong, John Shelby. *Living in Sin*. HarperSanFrancisco, A Division of HarperCollins Publishers, 1998, p. 136.
2. Ibid., pp. 111-112.
3. Spong, John Shelby. *Rescuing the Bible from Fundamentalism*. HarperSanFrancisco, A Division of HarperCollins Publishers, 1992, p. 33.
4. Ibid. pp. 16-17.
5. Spong, John Shelby. *A New Christianity for a New World*. HarperSanFrancisco, A Division of HarperCollins Publishers, 2001, p. 9.
6. Ibid., p. 23.
7. bid., p. 45.

(First published at chowk.com on September 28, 2004)

CHAPTER 11

ABDUS SALAM—THE MIRACLE
SCIENTIST OF PAKISTAN

In the conditions of modern life, the rule is absolute; the race
which does not value trained intelligence is doomed...Today, we
maintain ourselves, tomorrow science will have moved over yet
one more step and there will be no appeal from the judgment
which will be pronounced...on the uneducated (Alfred North
Whitehead).

Abdus Salam is known to be a devout Muslim, whose religion
does not occupy a separate compartment from his work and family
life. He once wrote, "The Holy Quran enjoins us to reflect on
the verities of Allah's created laws of nature; however that our
generation has been privileged to glimpse a part of the design is a
beauty and a grace for which I render thanks with a humble heart
(Miriam Lewis, see n.1).[2]

Abdus Salam was a rare phenomenon of nature. He was born in
1926 in Jhang, a nondescript backdrop in Pakistan's countryside and
rose to gain the distinction of winning not only the Nobel Prize in
Physics, in 1979, but also numerous other recognitions worldwide. In
this respect, he was nothing less than a miracle of nature.

His academic career as a student was star-studded. He set a new
university record in the matriculation examination when he was
only fourteen years old. He completed his MA at the Government
College, Lahore in 1946 and won a scholarship to study at St. John's

College, Cambridge University. He was only seventeen years old in his fourth year at Government College when he published his first paper. "It was an ingenious improvement on the solution of an algebraic problem discussed earlier by the Indian mathematical genius Srinivasa Ramanujan."[3]

He passed his BA (honors) at St. John's with a double First (wrangler) in mathematics and physics in 1949. He received Smith's Prize from Cambridge in 1950 for the most outstanding pre-doctoral contribution to physics. He obtained his PhD in theoretical physics in 1952.

He then returned to Pakistan to teach at his alma mater and continue research in his chosen field (n.2). Finding his line of work was largely unappreciated and encountering difficulties in pursuing a meaningful career, he returned to Cambridge to teach in 1954 in the position of lecturer. Commenting on his sad plight in Pakistan, Nigel Calder wrote, "...the academic climate in Pakistan was wrong; science was ignored not only by the intellectual leaders of the new nation, but also by the brightest students. Salam, simply, was intellectually lonely...I feared if I (Salam) stayed in Lahore my work would deteriorate. Then what use would I be to my country? Better to be a lecturer in Cambridge than a professor in Lahore."[4]

In 1957, he received an appointment from the Imperial College of Science and Technology, London, as a professor of theoretical physics, and his career took off to a fruitful fulfillment.

Salam made his name in the scientific world early on while he was still working for his PhD. He extended the application of the renormalization method that Schwinger, Feynman, Dyson, and independently Tomonaga had previously proposed and for which they were honored with the award of a Nobel Prize. For a fuller description of Salam's contribution (elimination of overlapping infinities), see n.3.

His next significant work was in regards to parity violation (destruction of symmetry). It was generally believed that the parity was conserved; the particles and their mirror images were as likely to be "left-handed" as "right handed." See n.4.

Salam hypothesized that "all neutrinos are left-handed," which called for violation of parity in the weak interactions. This was against the grain of contemporary conventional wisdom. Duff described Salam's visit to the "formidable" Wolfgang Pauli to try his hypothesis on him. According to him, "...he submitted (or should I say humbly submitted) his two-component neutrino idea. Pauli sent him packing unceremoniously with the jibe that the young man does not realize the sanctity of parity! So, Salam delayed publication until after Lee and Yang had conferred the mantle of respectability on parity violation."[5] After parity violation in the weak interactions had been experimentally verified by (Chien-Shiung) Wu's group and independently by Leon Lederman's group, Pauli wrote, "Now after the first shock is over, I begin to collect myself. Yes, it was very dramatic."[6]

The work that really distinguished Salam from his contemporaries was his unification of the electro-weak force for which he shared the 1979 Nobel Prize with Glashow and Weinberg. Tom Kibble, paying tribute to Salam wrote that Salam's work "...culminated in the discovery in 1967 of the electro-weak theory, showing how the electro-magnetic force responsible for most of chemistry and atomic physics, and the weak force manifested in radioactive decay, can be seen as part of a unified symmetric structure."[3]

To some people, Salam's genius was a kind of mystery. His approach to the problems that he worked on could be elusive and not self-evident. Duff wrote, "I think it was Hans Bethe who said that there are two kinds of genius. The first group (to which I would say Steven Weinberg belongs) produce results of such devastating logic and clarity that they leave you feeling that you could have done that too (if only you were smart enough). The second kind are the 'magicians' whose sources of inspiration are completely baffling. Salam, I believe, belonged to this magic circle and there was always an element of eastern mysticism in his ideas that left you wondering how to fathom his genius."[5] Similar views were expressed by the world-renowned Cambridge cosmologist, Fred Hoyle, also.

Founder of the International Centre for Theoretical Physics (ICTP)

Salam was acutely conscious of the backwardness in the field of science and technology of the Third World, particularly the Islamic world, to which he belonged. At the same time, he was profoundly proud of his glorious heritage also on which the Western world later erected the edifice of modern science. He also knew that merely commiserating on the plight of the Third-World countries was unrewarding; something concrete needed to be done to found a tradition of scientific research in the Third World. He used his influence at the United Nations to get help for founding a research center of physics in the Third World. He wanted to establish this center in Pakistan. According to Dr. Munir Ahmed Khan, "The IAEA (International Atomic Energy Agency) scientific advisory committee which included Nobel Laureate Rabi and Homi Bhaba unanimously opposed it. Privately Bhaba wanted this center to be at Bombay and offered Salam to join him. Salam refused."[7] The proposal for Pakistan fell through when President Ayub Khan's government could not come up with the one million dollars required as the seeding money. Eventually, the center went to Italy at Trieste, Salam being its founding director. Had it come to Pakistan, a very healthy tradition of scientific research would have been planted there.

Salam was very sensitive and painfully aware of the backwardness of the Muslim world in science and technology. Like any responsible and caring member of the ummah, he wanted to change all this. On one occasion, he wrote, "I can still recall a Nobel Prize Winner in Physics from a European country say this to me some years ago: Salam, do you really think we have an obligation to succor, aid, feed, and keep alive those nations who have never created or added an iota to man's stock of knowledge? And even if he had not said this, my own self-respect suffers a shattering hurt whenever I enter a hospital and reflect that almost every potent life-saving medicament of today, from penicillin upwards, has been created without our share of input from any of us from the Muslim world."[1] So, he wanted to change the existing situation for better by inviting budding scientists from the Third World and funding their stay at the ICTP to work not only on fundamental research but also on applied research. Many promising scientists benefited from Salam's center. But unless such a center and similar others in other areas of science and technology are created within the Third World itself, the real impact will not be felt.

You Can Always Tell the Truth

Duff described an interesting anecdote in his "Tribute to Salam" lecture, which reveals a glimpse of Salam's personal character. One of Salam's students was enmeshed in a dilemma in his work. He was getting two different sets of results from his calculations. One of these sets seemed irrelevant to him while he liked the other set. Faced with this problem, the student sought help from Salam. He said, "Professor Salam, these calculations confirm most of the arguments I have been making so far. Unfortunately, there are also these other calculations which do not quite seem to fit the picture. Should I also draw the reader's attention to these at the risk of spoiling the effect or should I wait? After all they will probably turn out to be irrelevant."

Salam said in response, "When all else fails, you can always tell the truth."

Poh'nchi Waheen Yeh Khak Jahan Ka Khameer Thha

(This dust returned to the place from where it had originated)

Professor Salam died on November 21, 1996. He had developed a degenerative neurological disorder that made his life increasingly difficult. According to Kibble, "He bore the affliction with remarkable stoicism, continuing to work so long as he was physically able, on new ideas both in theoretical physics and for third world development. He died peacefully at his home in Oxford. He had six children, four by his first wife and two by his second."[3]

Many friends, colleagues, and students paid tribute to their departed friend and mentor. Dr. Munir Ahmed Khan wrote, "My last meeting with Salam was only three months ago (August 1996). His disease had taken its toll and he was unable to talk. Yet he understood what was said. I told him about the celebration held in Pakistan on his seventieth birthday. He kept staring at me. He had risen above praise. As I rose to leave he pressed my hand to express his feelings as if he wanted to thank everyone who had said kind words about him."

Inna lillahay wa inna alaihay raj'aun.

Concluding his obituary, Dr. Munir Khan wrote, "Professor Salam had deep love for Pakistan in spite of the fact that he was treated unfairly and indifferently by his own country. It became more and more difficult for him to come to Pakistan and this hurt him deeply. Now he has returned home finally, to rest in peace for ever in the soil that he loved so much. Maybe in the years to come we will rise above our prejudice and own him and give him, after his death, what we could not when he was alive."

I hope so, too.

Honors and Awards

The world conferred numerous honors and awards on Professor Salam. Scores of universities from the Third World, Europe, and America gave him honorary D.Sc.'s. He was knighted by the Queen of the United Kingdom, and he received similar honors from several other countries as well.

He became a Fellow of the Royal Society (FRS), London, at the age of thirty-three. He was honored by the award of the Atoms for Peace, the Einstein medal, the J. Robert Oppenheimer Memorial medal, and the Hughes medal of the Royal Society, among many other honors. Pakistan honored him with the award of Sitara-e-Pakistan, the Pride of Performance medal, and the Order of Nishan-e-Imtiaz, which is the highest civilian honor. India honored him making him an honorary Fellow of the Tata Institute of Fundamental Research, Bombay. The ICTP renamed itself as the Abdus Salam International Center for Theoretical Physics after Salam's death, in his honor.

Indira Gandhi invited Salam to India when he received the Nobel Prize in 1979. He said he would visit Pakistan first. Pakistan invited him as a state guest. Dr. Munir Khan reported, "Once while visiting Beijing, I was told that the Chinese Academy hosted a dinner in his honor which was to be attended by the prime minister. However, breaking all protocol, the President also decided to attend the dinner just to honor Salam."[7]

He was scientific advisor to the government of Pakistan during President Ayub Khan and Yahya Khan's time. He continued in this position shortly in Zulfiqar Ali Bhutto's time, also.

I saw Professor Salam a few times at Imperial College when I was a student there (1968-70). I met him in his office for a few minutes for some personal business. In my own professional circles, I sensed that respect for me had increased a wee bit because Professor Salam was my compatriot.

(*Aasman teri lahd par shabnam afshani karay*)
(May the Heaven sprinkle dew on your grave)

Postscript—Excerpts from Professor Abdus Salam's Nobel Lecture, 8 December 1979

Seven hundreds and sixty years ago, a young Scotsman left his native glens to travel south to Toledo in Spain. His name was Michael, his goal to live and work at the Arab Universities of Toledo and Cordova, where the greatest Jewish scholar, Moses bin Maimoun, had taught a generation before.

Michael reached Toledo in 1217 AD. Once in Toledo, Michael formed the ambitious project of introducing Aristotle to Latin Europe, translating not from the original Greek, which he did not know, but from the Arabic translation then taught in Spain. From Toledo, Michael traveled to Sicily, to the Court of Emperor Frederick II.

Visiting the medical school at Salerno, chartered by Frederick in 1231, Michael met the Danish physician, Henrick Harpestraeng—later to become Court Physician of King Erik Plovpenning. Henrick had come to Salerno to compose his treatise on blood-letting. Henrick's sources were the medical canons of the great clinicians of Islam, Al-Razi and Avicenna, which only Michael could translate for him.

In respect of this cycle of scientific disparity, perhaps I can be more quantitative. George Sarton, in his monumental five-volume History of Science chose to divide his story of achievement in sciences into ages, each age lasting half a century. With each half century he associated one central figure. Thus 450 BC—400 BC, Sarton calls the Age of Plato, this is followed by half centuries of Aristotle, of Euclid, of Archimedes, and so on. From 600 AD to 650 AD is the Chinese half century of Hsiian Tsang, from 650 to 700 AD that of I-Chang, and then from 750 AD to 1100 AD—350 years continuously—it is the unbroken succession of the Ages of Jabir, Khwarizmi, Razi, Masudi, Wafa, Biruni and Avicenna and then Omar Khayam—Arabs, Turks, Afghans and Persians—men belonging to the culture of Islam. After 1100 appear the first Western names: Gerard Cremona, Roger Bacon—but the honors are still shared with the names of Ibn Rushd (Averroes), Moses Bin Maimoun, Tusi and Ibn Nafis—the man who anticipated Harvey's theory of circulation of blood.......

Notes

n.1. Miriam Lewis was on the staff of the ICTP. However, she was at IAEA when she wrote a short biography of Professor Salam.

n.2. "The head of Salam's institute (Government College?) told him that though he knew Salam had done some research, he could forget about it. He offered Salam a choice of three jobs: bursar, warden of a hall of residence, president of the football club. I (Salam) chose the football club." (This was probably additional faculty duty? author).

Excerpted from, "An Interview with Dr. Abdus Salam," *New Scientist*, 26 August 1976.

n3. The story of Salam's contribution to the renormalization theory is fully described by Robert P. Crease and Charles C. Mann in their book *The Second Creation*.[8] This rather detailed excerpt, which makes Salam narrate his contribution, is from that book.

I went to (Paul) Matthews and I said, "What have you been up to?" He said he had spent two and a half years trying to renormalize meson theories. He had found that only spin zero mesons would work. "This was encouraging: the pion has a spin of zero. "He had done the calculations to one-loop order and shown that the theory of spin zero was renormalizable up to the second order. Matthews said to me that I should read Dyson and look at the general problem of renormalization of meson theories. Following Dyson, in a couple of days I produced a general scheme of renormalizing all meson theories of spin zero.

That quickly?

"Very quickly," Salam said chuckling. "I went to Matthews and I said: Look, is this the problem?" He laughed and he said, "This is not the problem. You've done dimensional analysis, which only shows that various factors fit and everything would be fine—if one can show that the infinities really can be removed, each to its proper place. That's the problem. The problem devolves to a rather obscure point in Dyson's papers about overlapping infinities. It's so obscure we really have to work on that. "In complicated interactions, the reaction can go more than one way along the tangle of loops and vertexes to the graphs. The question was whether each subinfinity had to be removed individually, as if the others were not present. Dyson claimed to have solved this for quantum electrodynamics, but had not given his proof. In the meson theory, Matthews had encountered even more virulent snarls of infinities, many of which overlapped in a vicious manner. Would Dyson's claim hold for these also? Matthews said, "I'm taking vacation. You can have this problem until I come back to work in October. If you don't solve it till then, I'll take it back." That was the sort of gentleman's agreement we had. This was probably April or May, 1950.

Salam produced the solution before the stipulated time. Paul Matthews was incidentally Salam's supervisor at St. John's College and later a colleague at Imperial College.

n.4. Particles have a quantum mechanical property of parity. According to quantum mechanics, the conservation of this parity is equivalent to the laws of physics being invariant under mirror reflection. It has been found that parity is not conserved in the weak force and so the weak force is not invariant under mirror reflection.

http://www.en.wikipedia.org/Wiki/Parity

References

1. (Quoted by) Abdus Salam. "The Future of Science in Islamic Countries," part of a paper for inclusion in a volume presented to the Islamic Summit held in Kuwait, January 1987.
2. Lewis, Miriam. "Abdus Salam—Biography," http://www. nobel-se/physics/laureate/1979/salam-bio.html
3. Kibble, Tom. "Emeritus Professor Abdus Salam, Nobel Laureate," Staff Newspaper of Imperial College of Science, Technology and Medicine, 3-16 December, 1996, Obituary. Kibble was Salam's colleague and collaborator at Imperial.
4. Calder, Nigel. "A Man of Science—Abdus Salam. in *The World Treasury of Physics, Astronomy, and Mathematics.* Timothy Ferris, ed. Boston: Little, Brown, and Company, 1991, p. 670.
5. Duff, M.J. "A Tribute to Abdus Salam." An after-dinner talk delivered at the Workshop on Frontiers in Field Theory, Quantum Gravity, and String theory, Puri, India, 12-21 December, 1996. Duff was one of Salam's graduate students at Imperial College.
6. Myneni, Krishna. "Symmetry Destroyed: The Failure of Parity," http://www. ccreweb.org/documents/parity.html
7. Khan, Munir Ahmed. "Salam Passes Into History." The *News International*, Sunday, November 24, 1996, Page 7, Opinion.

(First published at chowk.com on July 26, 2004)

CHAPTER 12

RELIGION WITHOUT SCIENCE IS BLIND

Deliberating on spirituality and religion, Einstein asserted, "Science without religion is lame, religion without science is blind."[1] It is hard to understand this assertion in its literal sense, but from a broad viewpoint, it suggests that religion and science are not whole (complete) without each other. Religion is liable to run into blind alleys if it's not guided by science and if science is used without a spiritual aspect (religion), it might not make much sense either. It is known that Einstein, although of Jewish lineage, did not believe in any particular historical and divinely revealed religion. He did not speak of any particular religion when he made the preceding assertion. According to Bertrand Russell, "The word religion is used nowadays in a very loose sense. Some people under the influence of extreme Protestantism employ the word to denote any personal convictions as to morals or the nature of the universe. This use of the word is quite unhistorical."[2] When Einstein spoke of religion without specifying any particular historical religion, he spoke of religion in the abstract sense or in a philosophical context, which does not necessarily allude to any particular religion. At another place, he asserted, "My religion consists of a humble admiration of the illimitable superior spirit who reveals himself in the slight details we are able to perceive with our frail and feeble minds. That deeply emotional conviction of the presence of a superior reasoning power, which is revealed in the incomprehensible universe, forms my idea of God."[3] His God was not a god of retribution and reward because he also asserted,

"I cannot imagine a God who rewards and punishes the objects of his creation, whose purposes are modeled after our own—a God, in short, who is but a reflection of human frailty."[4]

When religion is viewed in this broad sense, it can be said that religion without science is blind because progressive scientific developments can broaden the religious perspective and make it more relevant to humankind. However, difficulties arise with the divine religions, which have Divine Scriptures that are believed to be the Word of God, as part of their creed. These traditional religions do not seek direction from science because they are believed to incorporate in them the "ultimate knowledge," which is derived from the divine revelations. There are thus unavoidable factual and intellectual conflicts, which exist between the divinely revealed religions and science. Pointing toward such conflicts, Einstein observed, "For example, a conflict arises when a religious community insists on the absolute truthfulness of all statements recorded in the Bible. This means an intervention on the part of religion into the sphere of science; this is where the struggle of the Church against the doctrines of Galileo and Darwin belongs."[5]

Nevertheless, these kinds of conflicts can be resolved if it is accepted that portions of the revealed scriptures are embodied in allegorical language which need not be read and understood literally. Such texts, which appear to conflict with the observed and scientific facts should be interpreted allegorically to reconcile the recondite scriptural meaning with the scientific truths. If such a resolution is not sought and accepted, religion and science will continue to be "blind" and "lame" in the Einsteinian sense, without any remedy.

It is not only the theologians who are seriously handicapped by adhering to the literal reading of the scriptures, the scientists are also liable to be guilty of similar culpability. According to Francisco Ayala, former Dominican priest and a professor of biological sciences at the University of California at Irvine, "Applying the criteria of scientific truth to religious claims is to make what philosophers call categorical mistakes...In a sonnet, Shakespeare may refer to his beloved as a rose. A scientist could say, 'This guy is an idiot. A woman is not a rose.' Of course, the idiot would be the one who made that comment. Shakespeare knows she is not a rose! But this doesn't mean that describing his beloved as a rose is not telling the world a lot about what he thinks about her, and what she is like, and what love is like."[6]

It appears that the Western world has reached a kind of rapprochement between religion and science. Science (and philosophy) and the church are separated from each other in the same way that state and the church are. Any conflicts arising between them are discussed intellectually, and reconciliation is sought for the resolution without resorting to harsh and violent means. An instance of such a conflict is the ongoing debate between the biological evolutionists and creationists. Also, there are several different degrees and gradations of religious belief and unbelief in Christianity. According to one such belief, "Science can no more answer the question of how we ought to live than religion can decree the age of the earth. Honorable and discernable scientists…have always understood that the limits to what science can answer also describe the power of its methods in their proper domain."[7] Another variety of beliefs is that of the evolutionists, who are also devout Christians. One of them is the Anglican priest Arthur Peacocke (see chapter 9) whom I had quoted in my earlier paper.[8] Peacocke and other religionists like him see "Darwinism as supporting his/her faith, not threatening it."[9]

Similar separation and symbiosis is desirable in other religions also if they are prone to run into conflict with the rational sciences. In Islam, conflicts of religion with science are rare these days because there are hardly any forefront scientists doing any meaningful science in the Muslim world. However conflicts exist between religion and rationalists. Religion can benefit if such conflicts are resolved peacefully through allegorical interpretation of the scriptural text which might appear to conflict with rational and empirical facts. Such a resolution was proposed early in the history of Islam by the first Arab philosopher al-Kindi (801-873 CE). His idea of allegorical interpretation was further worked upon by al-Farabi (870-950 CE) and formalized by Ibn Rushd (1128-1198 CE). Ibn Rushd's proposed method was called the doctrine of double truth.

Although there is the provision of ijtihad in Islam, it has not been used for the last one thousand years or so. Islam seems to have become static. The history of Islamic intellectual development after the Abbasid rule (749-1258 CE) was mostly dominated by the orthodox ulema, and the conflicts arising between religion and rational thought were allowed

119

to fester. No attempt was allowed to resolve such conflicts through reinterpretation of the scriptural text. The inevitable result was that nobody undertook to develop the rational and physical sciences. The Islamic philosophers are not philosophers in the true and general sense; the majority of them are religious metaphysicians. As an example, there is hardly any philosopher of science in the Islamic world. The Islamic philosophers are not generally well versed in physical sciences.

It is the need of modern times to resolve the conflicting issues peacefully through appropriate interpretation of the allegorical scriptures. That is the only way that religion can use science and rational thought positively for its own revitalization.

References

1. Albert Einstein's Words on Spirituality and Religion,http://www.deism/Einstein1.htm
2. Russell, Bertrand. Has Religion Made Useful Contributions to Civilization? http://positivecatholicism.org/hist/russell2.htm
3. See Ref. 1.
4. See Ref.1.
5. Albert Einstein on Science, Philosophy and Religion, http://www.update.uu.se-1bendz/library/ae_scire.htm
6. Quoted by Gordy Slack in "When Science and Religion Collide or Why Einstein Wasn't an Atheist?" http://www.motherjones.com/mother_jones/ND97/slack.html
7. Johnson, Philip E. *Darwin on Trial*. Regnery Gateway, Washington, D.C., 1991, p. 124.
8. Gill, Mohammad. "Conflict of Science with Theocracy," www.chowk.com, September 7, 2003.
9. Ruse, Michael. *Darwinism: Science or Philosophy*. Chapter 5, "Theism and Darwinism…" http://www.leaderu.com/orgs/fte/darwinism/chapter5.htm.

(First published at chowk.com on October 26, 2003)

CHAPTER 13

IS SCIENCE WITHOUT
RELIGION REALLY LAME

"Religion without science is blind and science without religion is lame" is one of the most popular sayings of Einstein on religion and science. I have already published an essay (on Chowk) on the first half of this saying, namely "Religion Without Science Is Blind."[1] (See chapter 12.) Religion indeed needs science to become more meaningful as the recent history of science shows. There had been conflicts between scientific truths and religious (Christian) beliefs in the past which were later resolved after religious views were appropriately adjusted in conformance with science. So in this perspective, it appears that religion needs science more than science needs it.

Does science need religion for its existence, wholeness (or even wholesomeness), or progress? A purview of the history of science shows that it was never dependent on religion at any period of time. Scientific discoveries had been made even before the inception of the organized religions. Since there is no single universal religion in the world, the question arises naturally: Which religion does science need to depend on, if it is really lame without it? It is true that the majority of the modern scientists believed in Judaism or Christianity, but the contributions of other scientists cannot be ignored. Many of the post-modern scientists are non-believers; they are agnostics or atheists. Einstein himself did not believe in any organized religion although his parents were Jews.

He was a believer in the sense that he believed in a non-personal God. He did not believe in a God of rewards and retributions. He was not agnostic like Bertrand Russell or atheist like his contemporary quantum physicist, Paul Dirac. Wolfgang Pauli said of Dirac, "Dirac has a new religion—there is no God and Dirac is His prophet."[2]

Einstein was not only a great, probably the greatest, physicist, but he was a great sage and a seer, also. Even if one takes a very broad reductionist view of the world religions and focuses on only the common denominator of all of them, i.e., the existence of God, it can be argued that science does not really depend upon such a thesis either. Consider the theories of relativity, special and general, for instance. They do not explicitly depend on God or belief in His existence. It may be true that Einstein may have been inspired by his faith in God during the development of his theories, but it is not quite the same thing as the inevitable need for believing in God. Scientists get inspirations from many other different sources, also.

For example, according to an anecdote described by Lederman, a physics Nobel laureate, Erwin Schrodinger received his inspiration romantically from his girlfriend. Lederman described this anecdote as follows:

> A few months after Heisenberg completed his matrix formulation, Erwin Schrodinger decided he needed a holiday. It was about ten days before Christmas in the winter of 1925. Schrodinger was a competent but undistinguished professor of physics at the University of Zurich, and all college teachers deserve a Christmas holiday. But this was no ordinary vacation. Leaving his wife at home, Schrodinger booked a villa in the Swiss Alps for two and a half weeks, taking with him his notebooks, two pearls, and an old Viennese girlfriend. Schrodinger's self-appointed mission was to save the patched-up, creaky quantum theory of the time. The Viennes-born physicist placed a pearl in each ear to screen out any distracting noises. Then he placed the girlfriend in bed for inspiration. Schrodinger had his work cut out for him. He had to create a new theory and keep the lady happy. Fortunately he was up to the task.[3]

Many scientists believe that even if God exists, He is on sabbatical now. According to them, after His act of creation in six days, God took His sabbatical for rest, and for good. His sabbatical is everlasting because His universe (a thing of beauty!) doesn't need any tweaking from Him; it is finely tuned, self-wound, and self-operational. The act of creation is still continuing without His intervention, in the sense that the universe is still unfolding and evolving physically. There are fundamental laws of nature which physics has discovered and the processes of biological evolution and natural selection which govern the evolution of the universe. The scientists are working hard to figure out how these laws actually operate. And in all this, nobody ever invokes God. It is true that many scientists believe in religion (different religions), and they find intuitive insights and inspirational succor from their individual religions, but they do not blend religion with their scientific theories. Professor Abdus Salam, after learning of his selection for the Nobel Prize, immediately offered namaz and gratitude to Allah. This must have given him a religious and moral uplift and recharged his batteries for continuation of his work in physics, but it did nothing to his science.

Science generally inculcates a sense of humility in the scientists who are overwhelmed by the realization of how little they know about nature and our universe. If humility is a divine quality, then in some sense, even agnostics and atheists are religious. Even in this narrow sense, humility only makes a scientist religious; science in itself remains untouched and uninfluenced.

I had briefly examined the relationship of science and religion in one of my earlier essays, in which I had written:

> There are Laplaces (non-believers) and Lagranges (believers) among the modern scientists, also, and God, per se, has still not been included in the scientific schemes of cosmology. There is no chapter on God in any standard textbook of physics. Apparently, there is simply no need. And there is not a single equation in physics which includes a parameter for God...Physics deals with mathematical relationships about the dynamics of the material objects

and their energies. God has not been described in sufficiently clear terms so that it is not possible to set up mathematical relationships which could describe explicitly how God plays a role in the universe. Nobody even knows if God has a material existence.[4]

If in some abstract and Einsteinian sense science is handicapped or crippled (lame) by the absence of religion, it is not clear how. On the other hand, many scientists believe that if religion is introduced into science, science becomes tainted. Einstein had great authority in science and because of his extremely high stature in science people had implicit faith in whatever he said.

But the scientists are a different breed. Their discipline requires them to question everything and everyone. A minor example of such a skeptical creed is Chandrasekhar's quip to Einstein's dictum, "God does not play dice with the world," which was, "How does he (Einstein) know that God does not play dice?" Paul Dirac questioned why Einstein was obsessed with God.

Einstein had a long controversy stretched over a period of thirty years with Niels Bohr. The central point of this controversy was the statistical character of quantum mechanics. Einstein had a firm belief that quantum mechanics ought to be deterministic like classical physics. It is stochastic only because we still do not know enough. In deterministic science, there wouldn't be any place for (Heisenberg's) indeterminacy principle. Niels Bohr and other quantum physicists believed indeterminacy was fundamental to quantum mechanics.

Einstein had significantly contributed to the development of quantum mechanics in its early stages, but he could not have easy symbiosis with the later developments. He in fact became quite irrelevant to quantum mechanics afterwards.

Likewise, he had problems in accepting the concept of the expanding universe, which was predicted by his theory of general relativity. Like many other physicists and astrophysicists in the first quarter of the twentieth century, he believed that the universe was static. He adhered tenaciously to this traditional view. To obtain such a universe from his theory, he had to arbitrarily fudge his equations by introducing a constant, which he called the "cosmological constant." Einstein did not accept the theoretical prediction of the expanding universe, which Friedmann had published in 1922 having deduced it from his (Einstein's) equations of general relativity. Einstein had speculated that Friedmann made a mistake in his calculations. However, Friedmann informed him that no such mistake existed in his work.

According to Stephen Hawking, "Only one man, it seems, was willing to take general relativity at face value. While Einstein and other physicists were looking for ways of avoiding general relativity's prediction of a non-static universe, the Russian physicist Alexander Friedmann instead set about explaining it."[5]

Afterwards, when Hubble's observations weighed in favor of an expanding universe, Einstein proclaimed his cosmological constant as the biggest blunder of his life. As it turned out, it is a blessing in disguise. The cosmological constant has acquired a different meaning in modern cosmology and is considered to be an essential ingredient in the potpourri of modern theories.

Thus science does not recognize any personal authority; it does not succumb to even the greatest scientist if his or her dictums and formulations are against the natural phenomena.

So in conclusion: Is science lame without religion? I do not believe so. It still has two legs (without any help from religion) if it was a biped to start with. Fortunately, it doesn't need any props; it stands and falls by its own standards. Science explains how natural phenomena occur; it does not always explain why they occur. Due to our ignorance, we need religion to explain "why." But such explanations may not be true, and they are not verifiable. Science does not accept anything until it is verified by empirical data. So we see that science and religion do not live in the same space.

However, science may be regarded as lame in the sense that its knowledge is limited. The moment one brings in religion to make it whole, it ceases to be science. It becomes metaphysics.

Conclusion

I would like to conclude this essay by giving hereunder an excerpt from Helge Kragh's book *Cosmology and Controversy* to show how (Georges Edouard) Lemaitre, a Belgian Catholic Priest and a prominent cosmologist of the first half of the twentieth century, kept his religion and science from getting into each other's way. Even though the expanding universe concept that he had presented was indicative of the possibility of the universe coming into existence at a finite time and thus the possibility of having been created, he resisted the temptation to project his religious belief into his science. The excerpt is as follows:

> As a Catholic priest, Lemaitre was, of course, aware that discussions about the beginning of the world could not, in the minds of most people, be separated from the question of God's creation of the world. He was at first inclined to include this aspect in his discussion, but then decided not to. In the typescript of the first note of March 1931, there is a paragraph reading: "I think that everyone who believes in a supreme being supporting every being and every acting, believes also that God is essentially hidden and may be glad to see how present physics provides a veil hiding the creation." The paragraph was crossed out by Lemaitre, not because it did not represent his conviction, but because he found it unwise to introduce God in his purportedly scientific sketch.[6]

The above quoted note was published in *Nature*.

References

1. Gill, Mohammad. "Religion without Science is Blind," www.chowk.com, October 26, 2003.
2. Crease, Robert P., and Charles C. Mann. "The Man Who Listened." in *The World Treasury of Physics, Astronomy, and Mathematics*. Timothy Ferris, ed. Boston Little, Brown, and Company, 1989, p. 67.
3. Lederman, Leon. *The God Particle*. New York: Houghton Mifflin Company, 1996, p. 167.
4. Gill, Mohammad. "The Theory of Everything," http.//www.secweb.org/asset.asp?AssetlD=61, 2001.
5. Hawking, Stephen W. *The Theory of Everything*. New Millennium Press, 2002, p. 25.
6. Kregh, Helge. *Cosmology and Controversy*. Princeton, NJ: Princeton University Press, 1996, pp. 48-49.

(First published at chowk.com on January 13, 2004)

CHAPTER 14

WHO IS A MUSLIM?

Recently, I read a book *Between Jihad and Salaam—Profiles in Islam* written by Joyce M. Davis in which she has published interviews with a number of prominent personalities of the Muslim world. The book contains seventeen profiles and interviews. The first of them is with Hassan-al-Turabi of Sudan. The book includes profiles and interviews of Khurshid Ahmad, Abida Hussain, and Muhammad Aslam Saleemi of Pakistan.

During her interview with Saleemi, an Ameer of Jama'at-i-Islami at that time (1997), Davis brought up the question about the definition of a Muslim. Saleemi responded, "There were two qualifications: that you believe in one God and that you believe Muhammad was a prophet."[1]

Davis said to Saleemi, "Although I consider myself a Christian, I believed in both those statements."

Saleemi's stony face softened, and his eyes glowed with apparent surprise and joy. "You are welcome," he said, beaming.

Davis, however, was somewhat skeptical and seemed to believe that much more in addition was needed for becoming a Muslim. This incident pricked my mind. This question as to who really is a Muslim must have cropped up numerous times in the history of Islam, but didn't it come up in the recent history of Islam in Pakistan? Yes, indeed, it did.

During the 1953 riots against Ahmadis (Qadianis), the various religious factions in Pakistan had demanded with one voice that the

131

Ahmadis be declared kafirs. A public court was constituted with Justice Muhammad Munir as president to investigate the causes of the uprising. In order to determine whether Ahmadis were indeed kafirs, the court had to first comprehend who, as a matter of fact, a Muslim is.

This question is discussed on page 215 of the Munir Report which was released in April 1954. It is appropriate to quote from it verbatim:

> The question, therefore, whether a person is or is not a Muslim will be of a fundamental importance, and it was for this reason that we asked most of the leading ulama to give their definition of a Muslim, the point being that if the ulama of the various sects believed the Ahmadis to be kafirs, they must have been quite clear in their minds not only about the grounds of such belief but also about the definition of a Muslim because the claim that a certain person or community is not within the pale of Islam implies on the part of the claimant an exact conception of what a Muslim is. The result of this part of the inquiry, however, has been anything but satisfactory, and if considerable confusion exists in the minds of our ulama on such a simple matter, one can easily imagine what the differences on more complicated matters will be...
>
> ...Keeping in view the several definitions given by the ulama, need we make any comment except that no two learned divines are agreed on this fundamental? If we attempt our own definition as each learned divine has done and that definition differs from that given by all others, we unanimously go out of the fold of Islam. And if we adopt the definition given by any one of the ulama, we remain Muslims according to the view of that alim but kafirs according to the definition of every one else..." (p.218).

Our (Muslims') tragedy is that when we talk to non-Muslims, we tell them that nothing can be simpler than Islam. Among ourselves, we cannot even agree on something as fundamental as the definition of a Muslim.

Ghulam Ahmad Parvez was a modernist and a learned and esteemed scholar of Islam who attempted to interpret the Quranic injunctions according to the modern times. He had a considerable following in the liberal circles. He wrote an article "Fatwas of Kufr' in his monthly magazine, *Tulu-i-Islam*, in the August 1969 issue. In this essay, he published the results of his comprehensive research and showed that all the prominent contemporary ulama had declared one another, individually, kafirs. This was a sad reflection on the state of affairs in the Islamic world.

Although Davis did consider the definition of a Muslim given by Saleemi as somewhat offhand, not necessarily complete, and given unprepared and on the spur of the moment, on second thought, what else is there?

I do hope that Saleemi himself sincerely believed in the definition that he gave to Davis.

References

1. Davis, Joyce, M. *Between Jihad and Salaam*—Profiles in Islam. St. Martin's Griffin, 1997, p. 285, 293-294.

(First published at chowk.com on February 29, 2004)